Climate and Global
Environmental Change

L. D. Danny Harvey

Department of Geography, University of Toronto, 100 St George Street, Toronto,
Canada M5S 3G3

PRENTICE HALL

An imprint of PEARSON EDUCATION

Harlow, England • London • New York • Reading, Massachusetts • San Francisco • Toronto • Don Mills, Ontario • Sydney •
Tokyo • Singapore • Hong Kong • Seoul • Taipei • Cape Town • Madrid • Mexico City • Amsterdam • Munich • Paris • Milan

Pearson Education Ltd
Edinburgh Gate
Harlow
Essex CM20 2JE
United Kingdom

and Associated Companies throughout the World.

Visit us on the World Wide Web at:
www.pearsoned-ema.com

First published 2000
Second impression 2000

ISBN 0582-32261-8

British Library Cataloguing-in-Publication Data
A catalogue record for this book is available from the British Library.

10 9 8 7 6 5 4 3 2 1
04 03 02 01 00

Typeset in Garamond 11/12pt by 30

Produced by Pearson Education Asia Pte Ltd.,
Printed in Singapore (KKP)

Contents

Understanding global environmental change: themes in physical geography

Around the season of All Hallows' Eve, 1994, Tom Spencer invited Olav Slaymaker to embark with him on the ambitious project of developing a book series on physical geography designed to come to grips, at an undergraduate student level, with the ongoing revolution in environmental geoscience. This revolution is the third phase in a sequence of intellectual revolutions initiated by global plate tectonics, reinforced by profound new insights into the complexity of the world-wide events of the late Quaternary and now focusing around the theme of global environmental change. Whether one believes in the influence of Hallowe'en over the lives of people or not, the challenge was accepted and we have learned that, not only is the project timely for physical geography as we understand it, but that many other geographers and geoscientists of various stripes and persuasions have encouraged us to proceed.

The series is intended for use by first- and second-year level geography and environmental science students in universities or colleges of higher education in the UK. It is also suitable as text material for second- or third-year level geography and environmental science students in Canada and the USA. Our main assumption is that the reader will have completed an introductory physical geography course.

Our motivation is discussed more fully in Chapters 1 and 8 of Volume 1, but for both of us there is an urgency in launching this series that derives at one level from a concern with re-establishing the historical legitimacy of the physical geographic tradition within the academy. We perceive, rightly or wrongly, a crisis of confidence among geographers over the traditional strength of the field in establishing linkages between the natural and social science traditions; its achievements in regional and global analysis and its preoccupation with the Earth as the home of people over Holocene and contemporary timescales. At a deeper level, and in common with many thoughtful persons of goodwill, we are still impressed by the recklessness with which the natural spheres of Earth continue to be exploited and believe passionately that we are accountable to subsequent generations for the way in which we respond to both biogeochemically and societally induced global environmental change.

The first volume in this series engages the most general trends and concepts associated with global environmental change and traces some of their connections with classical physical geographic writings. Subsequent volumes look in greater depth at each of the Earth's spheres and their interactions. We are grateful to the outstanding scientists who have committed their expertise and we acknowledge the initial work of Sally Wilkinson, then Publisher with the Longman Group, in enabling this project's commissioning. Subsequently, the sterling work of Tina Cadle (Development Editor, Sciences) and of Matthew Smith (Commissioning Editor) with Addison Wesley Longman, have commanded our respect and appreciation. We are reminded that All Saints' Day normally follows All Hallows' Eve at the stroke of midnight. Though this series will take a little longer to complete, it is our hope that, in its final form, it will make a small but perceptible contribution to the understanding that students of the next generation will have of the Earth. While there is no shortage of concern over the environment, there is an enormous challenge in converting this concern into meaningful action. We hope that these volumes will show that the physical geographer does have an important and distinctive role to play in the global environmental change debate. We are all stewards of the Earth; if through education we can ensure that the next generation's contribution to global sustainability will be more effective than that of our own then we will have achieved an important aim.

August 1, 1997
Olav Slaymaker, Vancouver
Tom Spencer, Cambridge

Volume preface

This book is a condensed version of *Global Warming: The Hard Science*, that was prepared to form part of the series *Geography and Environmental Change*. All of the information found in this book can also be found in the longer book, along with substantially greater analysis of some of the issues covered here.

The focus of both books is the science of regional- and global-scale climatic change due to human activities – change that is widely referred to as "global warming". As such, both books deal with our understanding of the processes involved in translating the emissions of gases that affect climate into changes in their atmospheric concentration, in translating concentration changes into changes in the heat balance, and in translating changes in the heat balance into changes in climate and in sea level. Our primary concern will be with the so-called greenhouse gases (GHGs), which have a warming effect on climate, and on suspended liquid or solid particles in the atmosphere – known as aerosols – which have both cooling and warming effects on climate. Human activities have significantly altered the worldwide concentration of a number of important GHGs, and have significantly altered the concentrations of aerosols over vast regions of both continents and oceans.

This book is intended as a relatively thorough introduction to the science of global warming that would be appropriate for second- or third-year undergraduate university courses. Science writers, high school science teachers, and scientifically literate members of the general public, business organizations, and environmental groups that are interested in the global warming issue will also find this book to be a valuable resource.

The book is divided into three parts. Part I provides background information on the workings of the climate system, on natural causes of climatic change, on the ways in which humans can influence climate, and on observed changes in the climate system during the past century, when both human and natural causes of climatic change are likely to have been important. The discussion of the climate system is centred around the major components of the system, the mass and energy flows that link them, and the physics of the so-called greenhouse effect. Part II begins with an introduction to the hierarchy of computer models that are used to study the climate system and to project the impacts of human activities on climate. This is followed by a discussion of basic principles concerning the operation of the greenhouse effect and of climatic feedbacks, and a summary of the effects of changes in the concentration of GHGs and aerosols on the heat balance of the Earth. Next, the processes by which emissions of GHGs and aerosols lead to changes in their concentrations are discussed, followed by a discussion of the main features of the expected steady state and time-varying climatic response to increasing GHG concentrations. Part II closes with a discussion of projected sea level rise. The emphasis is on the role of fundamental principles in assessing the overall credibility of such projections and in developing an appreciation of the risks of major changes in climate if unrestrained emissions of GHGs continue. A glossary can be found at the end of the book which defines all terms of a technical nature.

A number of other environmental issues are closely linked to the global warming issue. For example, emissions of CFCs (the chlorofluorocarbons) and related compounds are responsible for depletion of stratospheric ozone (O_3) and contribute to global warming, since these gases are powerful GHGs. The depletion of O_3 is of concern because of its role in shielding life on Earth from harmful ultraviolet radiation, but it has a cooling effect on climate. The buildup of O_3 in the lower atmosphere has a number of harmful effects on plants and animals, but also a warming effect on climate. Emissions of sulphur oxides from the burning of coal, the refining of crude oil, and the smelting of certain minerals contribute to rain acid and to local air pollution, but also have a marked regional cooling effect on climate. Loss of tropical rainforests is of great concern because of the massive species extinction that would result and the loss of valuable local ecosystem services (such as protection from floods and erosion), but the destruction of the rainforests also has implications for the global climate. All of these issues will be discussed in this book, but only to the extent that they influence climate.

The emphasis in this book is on global-scale rather than regional or local climatic change. There are two reasons for this. First, projections of regional-scale climatic change are not at all reliable, and may never be reliable. Second, this book attempts to rely as much as possible on arguments and lines of evidence that rest upon fundamental and well-established principles, that can be developed and understood based on intuitive reasoning and simple, back-of-the-envelope calculations, and that are independent of the details of any one computer simulation model. Global-scale and very broad regional-scale changes lend themselves nicely to this approach. For example, this book derives fairly strong arguments about how mid-latitude soil moisture in summer should respond in general to global warming, and discusses the competing factors that determine how strong the response will be, but one cannot say anything credible about the *magnitude* or even the *direction* of soil moisture changes at any given location. An important implication of this is that global warming can be viewed as a collective *risk* that merits a collective response among the nations of the world. There is little scientific (or moral) justification for individual nations to develop a policy response to the threat of global warming based on the anticipated climatic changes within their own jurisdiction alone.

This book and *Global Warming: The Hard Science* stop short of assessing the impacts of projected climatic change and associated sea level rise, as that would be worthy of a whole separate book in its own right, and introduces another layer of uncertainty (depending in part on how individuals and societies respond to the change in climate). However, the final chapter, in Part III, outlines the elements that need to be taken into account in developing a policy response to the global warming issues, and includes a discussion of the type of scientific information that can be used in the development of a policy response.

L.D. Danny Harvey, Toronto
February 1999

Acknowledgements

I would like to thank the following people for freely providing data in electronic form for use in this book: George Boer (Plate 7), Dave Etheridge (Figure 3.1), Laurent Fairhead (Plate 7), Marcel Fligge (Figure 9.2b), Vivian Gornitz (Table 4.4), Jim Hanson and Makiko Sato (Figures 6.3, 6.4 and Plate 9), Richard Houghton (Plate 1 and Figure 7.2), Tim Johns (Plate 7), Jeff Kiehl and Timothy Schneider (Plate 7), Judith Lean (Figure 9.2b), Toshinobu Machida (Figure 3.1), James Maslanik (Figure 4.6), Gerry Meehl (Plate 7), David Parker and Phil Jacobs (Plates 3 and 4), Alan Robock and Melissa Free (Figure 9.2c), Makiko Sato (Figure 9.2c), Matthias Schabel (Plate 9), Sharon Stammerjohn (Figure 4.7), Ron Stouffer (Plate 7), Ian Watterson (Plate 7), Tom Wigley (Figure 11.7).

Thanks are also due to Patrick Bonham who copy-edited the original manuscript for this book.

The publishers wish to thank the following for use of the material detailed below:
Figure 2.1: Trenbeth and Hoar, *Geophysical Research Letters*, 23, 57–60, 1996; Figure 4.6: Maslanik, Serreze and Barry, *Geophysical Research Letters*, 23, 1677–1680, 1996; Figures 4.9 and 4.10: Lansez, Nicholls, Gray and Avila, *Geophysical Research Letters*, 23, 1697–1700, 1996; Figure 4.11: Lambert, *Journal of Geophysical Research*, 101, 21319–21325, 1996; Figure 4.14: Marenco, Gouget, Pages and Karcher, *Journal of Geophysical Research*, 99, 16617–16632; Figure 5.1: Harvey and Schneider, *Journal of Geophysical Research*, 90, 2192 2205, 1985; Figure 5.3: Harvey and Huang, *Journal of Geophysical Research* 1999; Figure 5.4: Harvey, *Global Biogeochemical Cycles*, 3, 137–153, 1989; Figure 7.3a: Archer, *Global Biogeochemical Cycles*, 10, 159–174, 1996; Figure 7.3b: Archer, *Global Biogeochemical Cycles*, 10, 511–526, 1996; © by the American Geophysical Union.
Figure 2.5: Sun and Linzden (1993) Distribution of tropical tropospheric water vapor, *Journal of the Atmospheric Sciences*, 50, 1643–1660, © 1993 AMS. Figure 3.3: Reprinted with permission from *Nature*. Dlugokencky *et al.*, (1998) 393, 447–45 0; Figure 4.1: Folland & Parker (1995) Correction of instrumental bias in historical sea surface temperature data. *Quarterly Journal of the Royal Meterological Society*, 121, 319–367.

Figure 4.7: *Climatic Change*, 37, 1997, 617–639, Opposing Southern Ocean climate patterns as revealed by trends in regional sea ice coverage, Stammerjohn & Smith, figure 3, with kind permission of Kluwer Academic Publishers.

Figures 4.5 and 4.8: Nicholls, Gruza, Jouzel, Karl, Ogallo and Parker (1996) Observed climate variability and change in Houghton, Filho, Callander, Harris, Kattenberg and Maskell (eds) *Climate Change 1995: The Science of Climate Change* (Cambridge: CUP); Figure 10.1: Warrick, Le Provost, Meier, Oerlemans and Woodworth (1996) Changes in sea level, in Houghton, Filho, Callander, Harris, Kattenberg and Maskell (eds) *Climate Change 1995: The Science of Climate Change* (Cambridge: CUP) © IPCC 1996.

Figure 8.3: Mitchell, Manabe, Melshko and Tokioka (1990) Equilibrium climate change and its implications for the future, in Houghton, Jenkins and Ephraums (eds) *Climate Change: The IPCC Scientific Assessment* (Cambridge: CUP) © IPCC 1990.

Figure 4.3: Jones, Osborn and Briffa (1997) Estimating sampling errors in large-scale temperature averages, *Journal of Climate*, 10, 2548-2568, © 1997 AMS.

Figure 7.1: Reprinted from *Agricultural and Forest Meteorology*, 69, Idso and Idso, Plant responses to atmospheric CO_2 enrichment in the face of environmental constraints: a review of the past 10 year's research, 155–203, © 1994; Figure 8.4: Reprinted from *Paleogeography, Paleoclimatology, Paleoecology*, 121, Kheshgi and Lapenis, Estimating the accuracy of Russian paleotemperature reconstruction, 221–237, © 1996 with kind permission of Elsevier Science.

Figure 5.2: *Climate Dynamics*, Climate sensititvity due to increased carbon dioxide: experiments with a coupled atmosphere and ocean general circulation model, Washington and Meehl, 4, 1–38, figure 1, 1989 © Springer-Verlag.

Figure 9.1: *Climate Dynamics*, A zonally averaged three-basin ocean circulation model for climate studies, Hovine and Fichet, 10, 313–331, figure 3a–d, 1994 © Springer-Verlag.

Figure 9.3: Bryant, *Climate Process and Change*, 1997, Cambridge University Press.

Figure 10.2 with permission from *Nature*. Oppenheimer (1998) 393, 325–332; Figures 11.6 and 11.7: Reprinted with permission from *Nature*, Hoffert *et al.*, (1998) 395, 881–884 © 1998 Macmillan Magazines Ltd.

Figure 10.3: *Climate Dynamics*, The steric component of a sea level rise associated with enhanced greenhouse warming: a model study, Bryan, 12, 545–555, figure 1a and 3b, 1996 © Springer-Verlag.

Though every effort has been made to trace the owners of copyright material, in a few cases this has proved impossible and we take this opportunity to apologise to any copyright holders whose rights may have been unwittingly infringed.

Part 1

Introduction

Chapter 1

Climatic change and variability – past, present, and future

Evidence from the geological record indicates that the Earth's climate has changed throughout the Earth's geological history, spanning more than 3 billion years. Geological evidence can also be used to reconstruct past variations in the concentration of CO_2 (carbon dioxide) and CH_4 (methane), which are two important heat-trapping gases. The ice preserved in the Antarctic and Greenland ice caps contains air bubbles that were sealed off within a few hundred years of the accumulation of the snow, thereby providing samples of the atmosphere for the last 250,000 years. From these air bubbles we can measure changes in the concentrations of CO_2 and CH_4. Other, less certain evidence has been used to reconstruct the variation in the atmospheric CO_2 concentration as far back as 570 million years (e.g. Berner, 1994). Past variations in the Earth's climate and in the composition of the atmosphere provide a long-term perspective against which the human-induced changes in the atmosphere, and projected changes in the atmosphere and in climate, can be compared. In this chapter this long-term perspective is presented.

1.1 A geological perspective

Figure 1.1 shows the reconstructed variation in the average surface air temperature of the Earth, with each successive panel showing greater detail for time spans progressively closer to the present. Figure 1.1(a) gives a generalized temperature and precipitation history of the Earth during the past 3 billion years based on a variety of paleoclimatic indicators, some of them qualitative in nature. Figure 1.1(b) shows the variation in the oxygen isotope composition of bottom-dwelling foraminifera in the global ocean during the past 70 million years, along with a rough quantitative temperature scale. The deep waters of the world's oceans decreased from about 12°C around 60 million years ago, to near 0°C at present. Figure 1.1(c) shows the variation in the oxygen isotope composition of bottom-dwelling foraminifera at Ocean Drilling Project site 677 during the past 1.2 million years; fluctuations at this time scale are largely the result of variations in the global volume of ice on land. Figure 1.1(d) shows the variation in the oxygen isotope composition of ice that accumulated on Devon Island, in the Canadian Arctic Archipelago, during the last 120,000 years. These fluctuations are related to changes in the temperature of the air masses over the ice, and a rough temperature scale is provided based on the

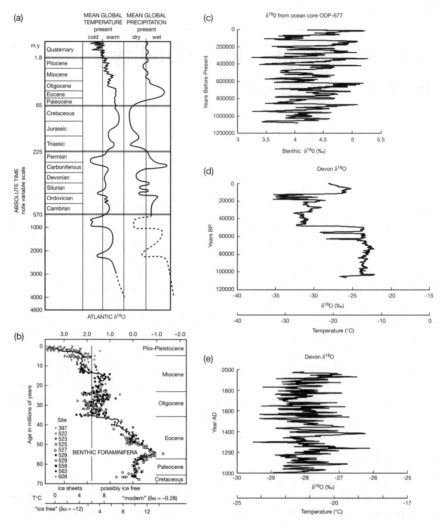

Figure 1.1 (a) Reconstructed variation in global mean surface air temperature and precipitation during the past 3.0 billion years, from Frakes (1979). (b) Variation in the oxygen isotopic composition of deep-sea foraminifera in the global ocean during the past 70 million years, and an approximate interpretation in terms of the temperature of deep ocean water, from Miller *et al.* (1987). (c) Oxygen isotope variation of deep ocean sediments at ODP site 677 during the past 1.2 million years. (d, e) Variation in the oxygen isotope composition of ice on Devon Island during (d) the past 120,000 years, and (e) the past 1000 years, along with a very rough scale for temperature fluctuations. Data for (c) were obtained from the paleoclimatic data web site: ftp://ftp.ngdc.noaa.gov/paleo/paleocean/sedimentfiles/complete/odp-677.csv, while data for (d) and (e) were obtained from ftp://ftp.ngdc.noaa.gov/paleo/polar/devon/devon data/d72del.200, and ftp://ftp.ngdc.noaa.gov/paleo/polar/devon/devon data/d7273del.5yr, respectively.

correlation between the oxygen isotope composition of Arctic precipitation and temperature given in Johnsen *et al.* (1989). Finally, Figure 1.1(e) shows the variation in the isotope ratio in Devon Island ice during the past 1000 years. From these data it can be seen that the Earth's climate changes at a wide range of time scales, with decadal-scale variations superimposed on century and millennial time-scale variations, which in turn are superimposed on variations spanning tens of thousands of years, which are superimposed on still longer time-scale variations, and so on. Major episodes of glaciation, interspersed with interglacial intervals, occurred during the pre-Cambrian period, during the Permo-Carboniferous period, and during the past 3 million years. In between these glacial–interglacial episodes have been extended periods with climates substantially warmer than during the present interglacial period. In short, the climate is a dynamic, constantly changing phenomenon.

Figure 1.2 shows the variation in atmospheric CO_2 concentration over the last 570 million years as inferred from computer models that simulate the geological time-scale flows of carbon. There is no direct record of atmospheric CO_2 concentration this far back, but measurements of quantities that would have been affected by changes in the carbon cycle, such as the chemical composition of oceanic sediments, can be used as a rough check on these models. According to the results shown in Figure 1.2, the atmospheric CO_2 concentration has varied significantly over periods of tens to hundreds of millions of years, reaching concentrations of 4–16 times the pre-industrial concentration of 280 ppmv (parts per million by volume), albeit with a large uncertainty range. Figure 1.3 shows the variation in atmospheric CO_2 and CH_4 concentration over the past 160,000 years as directly measured in air bubbles trapped in Antarctic ice. Also shown on Figure 1.3 are the present concentrations of CO_2

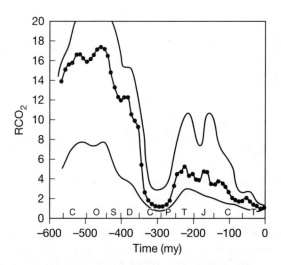

Figure 1.2 Reconstructed variation in atmospheric CO_2 concentration during the past 570 million years. Reproduced from Berner (1994).

and CH_4, and a continuation of the upward trend during the next few centuries as a result of human emissions. During the last 160,000 years, CO_2 underwent natural variations in concentration from as low as 180 ppmv to as high as 300 ppmv, while CH_4 varied between 0.3 and 0.7 ppmv. In contrast, human activities during the past 200 years have increased the CO_2 concentration to over 360 ppmv and the CH_4 concentration to over 1.7 ppmv. Furthermore, the rates of change during the last 200 years far exceed the rates of change that occurred naturally during the preceding 160,000 years.

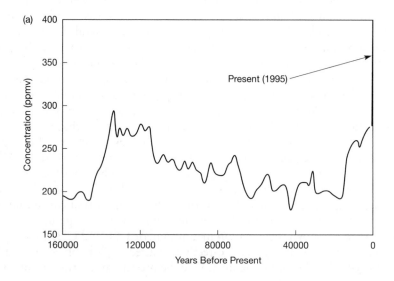

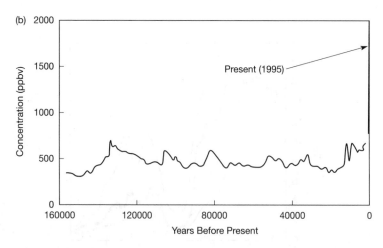

Figure 1.3 Variation in atmospheric (a) CO_2 and (b) CH_4 concentration during the past 160,000 years, as measured in air bubbles trapped in Antarctic ice. Based on data in Barnola *et al.* (1991), for CO_2, and Chapellaz *et al.* (1990), for CH_4. Both datasets were obtained from the National Oceanographic and Atmospheric Administration (NOAA) website, http://www.ngdc.noaa.gov/paleo.

1.2 Future prospects and the scientific basis of concern

Not only are the recent, human-induced changes in the atmospheric concentration of CO_2 and CH_4 unprecedented in speed and magnitude, but "business-as-usual" (BAU) projections (i.e., assuming no effort to restrain emissions) indicate that far larger changes will occur during the 21st century and beyond. Figure 1.4 shows a closeup of the BAU projections of atmospheric CO_2 and CH_4 concentration that were shown as a sharp spike at the

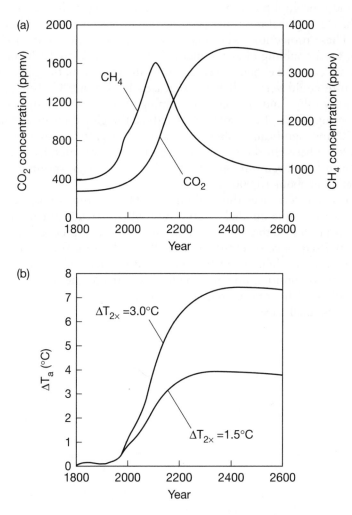

Figure 1.4 (a) Atmospheric CO_2 concentration (ppmv, left scale) and CH_4 (ppbv, right scale) as projected to the year 2100 according to a plausible business-as-usual emission scenario. (b) Change in global mean surface air temperature corresponding to the business-as-usual CO_2 and CH_4 concentrations shown in (a). Results in (b) are shown assuming eventual global mean temperature increases for a doubling of atmospheric CO_2 concentration of 1.5°C and 3.0°C.

right-hand edge of Figure 1.3. The emissions assumed for this projection are discussed in Chapter 4 but are not at all excessive. As seen from this figure, the atmospheric CO_2 concentration could reach six times the pre-industrial concentration, while CH_4 could reach four times the pre-industrial concentration. Significant increases in the concentrations of other GHGs are also projected. On the basis of verifiable laboratory measurements, these concentration increases will lead to a significant trapping of heat, and on the basis of very fundamental physical principles and observational evidence, this heat trapping will almost certainly lead to changes in climate that are significant from a human and ecological point of view. Also shown in Figure 1.4 are projections of the change in global mean temperature due to business-as-usual buildup of GHGs. These projections were made assuming that the eventual climatic warming for a doubling of atmospheric CO_2 only ($\Delta T_{2\times}$) is 1.5°C or 3.0°C, which is not excessive either. Nevertheless, by the year 2100, the global average temperature could very well have increased by 2.5–4.0°C above the late 19th century value, and an eventual warming of 3.5–7.0°C occurs. Inclusion of the cooling effect of aerosols might reduce the global mean warming by 0.5–1.0°C and significantly alter regional weather patterns. By comparison, the difference between the last ice age and the present interglacial period is estimated to have been only 4–5°C, but the post-glacial warming occurred over a period of about 10,000 years.

Scientific concern over the buildup of GHGs in the atmosphere is not based on some simple-minded extrapolation of past trends in concentrations or in temperatures. Rather, it is based on a step-by-step, process-based assessment of the sequence of events and feedbacks leading from emissions of GHGs to climatic change and the associated impacts. The changes in the atmosphere's composition provoked by human activities are unprecedented in recent geological history, extremely rapid, and – in the case of the CO_2 buildup – essentially irreversible.

Chapter 2

The climate system and climatic change

The climate in any given region involves both the average of the weather and the typical extent to which conditions fluctuate from the average. The term "weather" refers to the day-to-day changes in the state of the atmosphere at a specific location. It includes variables such as temperature, humidity, windiness, cloudiness, and precipitation. Although the climate is usually thought of in terms of atmospheric variables, it depends on much more than just the atmosphere. Rather, it depends on all of the components that interact together to form part of the *climate system*. In this chapter the major components of the climate system and the linkages and interactions between the components are described. An important point that emerges from this discussion is that a full understanding of climate requires linking together information – often quite specialized – from a wide range of disciplines. These disciplines include meteorology, physical and chemical oceanography, atmospheric and soil chemistry, cloud and aerosol physics, and marine and terrestrial ecology.

2.1 Components of the climate system

A system can be defined as a set of components such that each component influences, and is influenced by, all the other components. The climate system consists of the atmosphere, oceans, biosphere, cryosphere (ice and snow), and lithosphere (Earth's crust). Each of these components influences, and is influenced by, all of the others, so they form part of a single system. The sun is not part of the climate system because the climate cannot affect the sun. Rather, the sun is said to be an *external forcing*. Volcanic eruptions, which inject sulphur gases into the atmosphere that ultimately have a cooling effect, are also an external forcing because they are external in a system sense – they influence, but are not influenced by, the climate system.

The components of the climate system are linked by flows of energy and matter. The energy flows occur as solar and infrared radiation, as sensible heat (heat that can be directly felt or sensed), as latent heat (related to the evaporation and condensation of water vapour, or the freezing and melting of ice), and through the transfer of momentum between the atmosphere and ocean. The major mass flows involve water, carbon, sulphur, and nutrients such as phosphorus (P) and nitrate (NO_3^-). The behaviour of the climate system depends on how these energy and mass flows change as the system changes, on how the flows themselves influence the system, and on the speed with which the system responds to changes in the mass and energy flows. The main features of the energy and mass flows in the climate system are described next.

2.2 Energy flows

The major energy flows within the climate system are illustrated in Figure 2.1. Energy from the sun (solar energy) occurs at a variety of wavelengths, ranging from ultraviolet (0.1–0.4 µm) and visible (0.4–0.7 µm) to near infrared (0.7–4.0 µm). Much of the ultraviolet radiation is absorbed by ozone (O_3) in the stratosphere and, to a lesser extent, in the troposphere (the region from the surface to the base of the stratosphere). Some (a few percent) of the visible and near infrared radiation is reflected back to space by aerosol particles, and an even smaller fraction is absorbed by aerosols. The major types of aerosol are sea salts (which form particles when spray from the oceans evaporates), dust, soot (black carbon) aerosols directly emitted from the burning of biomass, organic carbon aerosols that are either produced by photochemical reactions involving gases emitted by plants (e.g., terpenes) or directly emitted, and sulphate particles that form from emissions of sulphur compounds by bacteria in the ocean, land plants, soils, and volcanoes. Clouds exert an enormous influence on the flow of solar energy, by typically reflecting 40–80% of the incident radiation back to space and by absorbing 5–15%.

The land surface, oceans, clouds, and atmosphere emit their own radiation, in the infrared part of the spectrum (wavelengths from 4.0 to 50.0 µm). The atmosphere absorbs a portion of the infrared radiation emitted from the surface, and replaces the absorbed emission with its own emission. Since the

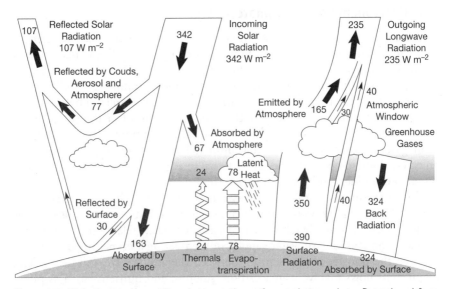

Figure 2.1 Global mean energy flows between the surface and atmosphere. Reproduced from Trenberth *et al.* (1996).

atmosphere is generally colder than the Earth's surface, the amount of infrared radiation that the atmosphere re-emits is smaller than the amount it absorbs, as seen in Figure 2.1. The net result is to reduce the amount of infrared radiation that escapes to space. The atmosphere also emits some infrared radiation back to the Earth's surface.

Since radiation – whether solar or infrared – is a form of energy, the reduction in the loss of infrared radiation to space by the atmosphere tends to make the climate warmer than it would be otherwise. This is the so-called "greenhouse effect" although, as explained in Box 2.1, the term is a misnomer since a greenhouse is warm largely for different reasons. The main gases responsible for the greenhouse effect are those gases that are capable of absorbing and re-emitting infrared radiation. The most important of these is water vapour, followed by CO_2 (carbon dioxide), O_3, CH_4 (methane), and N_2O (nitrous oxide). These are all naturally occurring gases, so the greenhouse effect is a naturally occurring phenomenon. Clouds are also effective greenhouse agents, as they absorb all of the radiation emitted from the Earth's surface below (except for thin, wispy clouds that are known as cirrus clouds) but re-emit a considerably smaller amount of radiation owing to the fact that the tops of clouds are generally much colder than the Earth's surface.

Box 2.1 The greenhouse effect and real greenhouses

The term "greenhouse effect" is used to refer to the tendency of the atmosphere to create a warmer climate than would otherwise be the case. However, the physical mechanisms by which the presence of the atmosphere warms the climate and the primary mechanism that causes a greenhouse to be warm are in fact quite different. A greenhouse heats up by day as the air within the greenhouse is heated by the sun. Outside the greenhouse, near-surface air that is heated through absorption of solar radiation by the ground surface is free to rise and be replaced with colder air from above. This cannot happen in a greenhouse – the heated air is physically prevented from rising and being replaced with colder air. The so-called greenhouse effect does not involve preventing the physical movement of air parcels. Rather, it involves the net trapping of infrared radiation, which occurs independently of the movement of individual air parcels or the lack thereof. There is, nevertheless, a weak similarity between the greenhouse effect and what happens in a greenhouse: the net trapping of infrared radiation occurs because the atmosphere absorbs part of the radiation emitted from the Earth's surface, and then re-emits a smaller stream of radiation owing to the fact that the atmosphere is colder than the Earth's surface. The glass enclosing a greenhouse will absorb close to 100% of the radiation emitted from within the greenhouse, but will re-emit a smaller amount of radiation to the sky if the outer surface of the glass is colder than the interior of the greenhouse.

Also shown in Figure 2.1 are the flows of sensible and latent heat from the surface to the atmosphere. The Earth's surface on average absorbs about 100 W m^{-2} more solar + infrared radiation than its emission of infrared radiation, while the atmosphere emits about 100 W m^{-2} more radiation than it absorbs. About 80% of the excess radiant energy at the surface is used to evaporate water rather than raising the surface temperature. When the water vapour condenses in the atmosphere, enough heat is released to offset about 80% of the cooling that would otherwise occur. This is referred to as a *latent heat* transfer. The remaining 20% is accounted for by rising warm air and sinking cold air – the *sensible heat* flux.

Heat is also transferred horizontally – as sensible and latent heat in the atmosphere, and predominantly as sensible heat in the oceans. These transfers cause the equatorial regions to be cooler than they would otherwise be (by removing excess radiative energy) and cause the polar regions to be warmer than they would otherwise be. Changes in the rate of horizontal heat transfer can be an important source of regional climatic change.

Important heat transfers also occur between the surface layer and deeper layers of the ocean. These transfers occur in three ways (Figure 2.2): through turbulent mixing or *diffusion*, which always transfers heat from warm to cold layers; when surface water becomes cold and denser than the underlying water and thereby switches places with the underlying water – a process called *convection*, which almost always transfers heat upward; and through the large-scale transfer of water from one place to another (*advection*). Cold water sinks in polar regions and spreads throughout the world ocean, causing the deep ocean to be cold (0–2°C). This sinking is balanced by the upwelling of relatively warm water in low and middle latitudes. This large-scale overturning is driven

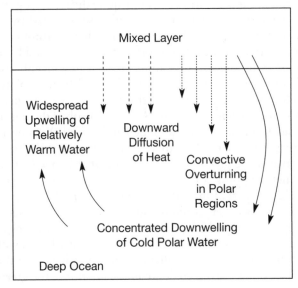

Figure 2.2 Global mean energy flows between the ocean surface layer and the deep ocean.

by horizontal variations in the density of water. The density of seawater in turn depends on its temperature and salinity, so the density-driven overturning is referred to as the *thermohaline overturning*. Although the water that upwells as part of the thermohaline overturning is colder than the surface water above it (and therefore cools the surface in the region where it upwells), it is warmer than the water that sinks, so the net effect is an upward heat transfer. However, because the ocean temperature decreases with increasing depth at most places, diffusive mixing results in a downward heat transfer.

For an unchanging climate, the net heat flow to or from the deep ocean must be zero. However, if the surface temperature warms, a net heat flow into the deep ocean will arise which will tend to slow down the subsequent warming until the deep ocean has warmed up – a process that takes thousands of years to complete. Conversely, if the intensity of sinking and upwelling were to change, there would be a temporary net flow of heat between the deep ocean and surface which would drive a change in the surface climate. In particular, a reduction in overturning would lead to a temporary surface cooling because the upward heat transfer caused by overturning would be reduced.

2.3 The hydrological cycle

The hydrological cycle involves the evaporation and precipitation of water, transport of water vapour by the atmosphere, runoff from the continents to the oceans, and the net transfer of freshwater by ocean currents. It is illustrated in Figure 2.3. As noted in Section 2.2, evaporation and precipitation account for some 80% of the required non-radiative energy transfer from the Earth's

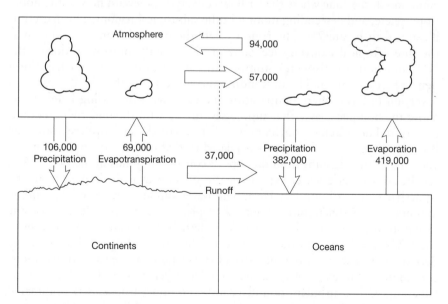

Figure 2.3 The global hydrological cycle, showing fluxes in Gt (billions of tonnes) of H_2O. Flux estimates were derived from Christopherson (1992, his Figure 7-10).

surface to the atmosphere. To the extent that condensation and the associated release of latent heat occur in different regions from where evaporation occurred, the hydrological cycle also effects important horizontal energy transfers. Geographical variations in the difference between precipitation + runoff onto and evaporation from the oceans are responsible for creating regional variations in surface salinity which, as noted in Section 2.2, contribute to driving (or opposing) the ocean's thermohaline circulation. Finally, water vapour is the single most important GHG, and both its amount and distribution within the atmosphere are of critical importance to climate.

The entire amount of water vapour in the atmosphere is completely replaced by evaporation and precipitation once every eight days. Changes in the amount of water vapour in the atmosphere will respond to changes in the rate of evaporation on a comparable time scale. Since the evaporation rate is directly related to surface temperature, it follows that the amount of water vapour in the atmosphere responds almost instantaneously to changes in the climate.

However, the quantitative link between climate and the amount of water vapour in the atmosphere depends on the processes that govern the vertical distribution of water vapour in the atmosphere. These are different at middle-to-high and at low latitudes. At middle and high latitudes, water vapour is transferred vertically by large-scale winds, associated primarily with travelling storm systems. The humidity in the free troposphere is closely coupled to that of the atmospheric layer next to the Earth's surface (the so-called boundary layer). At low latitudes (30°S–30°N), the atmospheric circulation is dominated by the Hadley cells. The Hadley cells consist of a flow of surface air toward the equator (or thereabouts) in both hemispheres, rising motion with intense precipitation in the zone where the air flow converges, poleward flow in the upper troposphere, and descending motion in the subtropical regions. This pattern is illustrated in Figure 2.4. The region where the two cells converge, known as the Inter-Tropical Convergence Zone or ITCZ, shifts northward during the NH (northern hemisphere) summer and southward in the SH (southern hemisphere) summer. The descending motion outside the ITCZ creates an inversion (a layer where temperature increases with increasing height) just above the boundary layer that suppresses mixing with the underlying boundary layer. The water vapour content of much of the atmosphere above the boundary layer in the subtropics is related to that of the very cold and hence very dry air that detrains from the tops of convective clouds in the ITCZ. As a result, atmospheric water vapour in the subtropics is decoupled from the underlying, moist boundary layer. The water vapour content of the middle and upper troposphere throughout the tropics and subtropics depends on the extent to which ice crystals in the air that detrains from convective columns in the ITCZ sublimate and thereby moisten the surrounding air. The extent of detrainment varies with the intensity of convection and hence with the moisture content and temperature of the boundary layer in the ITCZ. This vertical zonation, with a turbulent boundary layer, an upper troposphere dominated by detrainment from convective columns, and a middle troposphere dominated by subsiding air, is illustrated in Figure 2.5.

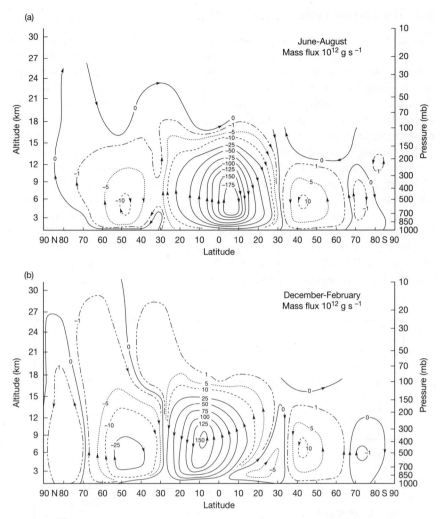

Figure 2.4 The zonally averaged flow associated with the Hadley cells in (a) NH summer, and (b) SH summer. Redrafted from Newell (1978).

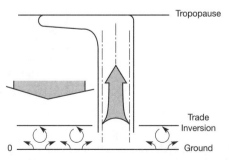

Figure 2.5 Water transfer processes associated with cumulus convection. Reproduced from Sun and Lindzen (1993).

2.4 The carbon cycle

The carbon cycle is critically important to climate because it regulates the amount of two important greenhouse gases in the atmosphere: CO_2 and CH_4. Carbon, like water, continuously cycles between various "reservoirs" or temporary holding areas, but this cycling is considerably more complicated for CO_2 than for water. Furthermore, whereas the amount of water vapour in the atmosphere is directly and instantly driven by changes in the climate, the CO_2 concentration in the atmosphere responds to changes in the climate on time scales ranging from a few months to thousands of years. This introduces the potential for long-term feedbacks between climate and the carbon cycle.

Figure 2.6 illustrates the major carbon reservoirs as a series of interconnected boxes. The number in each box is the estimated amount of carbon in the reservoir in units of gigatonnes (Gt, or billions of tonnes) prior to human disturbance, while the numbers beside the boxes are the annual rates of flow (or flux) of carbon from one reservoir to another. Carbon occurs in the atmosphere primarily as CO_2 and to a much smaller extent as CH_4; in the living biota and soils as organic matter; and in the oceans primarily as dissolved CO_2,

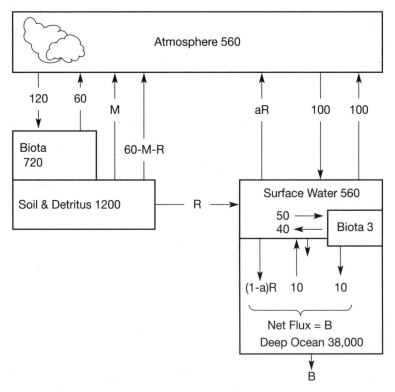

Figure 2.6 The global carbon cycle. Numbers in boxes are the amounts of carbon in each reservoir as Gt C, while numbers next to arrows between boxes are the annual rates of carbon transfer in Gt C. M = methane flux, R = riverine flux of particulate carbon and B = burial flux.

HCO_3^- (bicarbonate ion), and CO_3^{2-} (carbonate ion). The CO_2, HCO_3^-, and CO_3^{2-} are collectively referred to as dissolved inorganic carbon (DIC). The ocean can be divided into a surface layer and the remaining subsurface water. The surface layer is referred to as the *mixed layer* because it is well mixed by winds and the effects of seasonal cooling. Water parcels in the mixed layer are frequently brought to the ocean–air interface and can exchange CO_2 with the atmosphere.

From Figure 2.6 it can be seen that (i) the amount of carbon in the land biota is roughly comparable to the amount of carbon in the atmosphere; (ii) the amount of carbon in the soil and detritus is about twice the amount in either the atmosphere or above-ground land biota; (iii) the amount of carbon in the ocean mixed layer, which interacts directly with the atmosphere, is comparable to the amount of carbon in the atmosphere itself; and (iv) the overwhelming majority of carbon in the biosphere + atmosphere + ocean system is in the deep ocean. These facts have important consequences concerning the response of the carbon cycle to human emissions of CO_2, the ways in which climatic change itself can influence the carbon cycle, and the potential role of reforestation in offsetting fossil fuel emissions of CO_2.

Carbon dioxide is incorporated into living plant material through photosynthesis and released to the atmosphere through respiration (the process by which organic matter is "burned" to provide energy, either by the plant itself, or by organisms that consume and thereby decompose organic litter). Some of the biospheric carbon is returned to the atmosphere as CH_4 rather than CO_2, mainly from wetlands. The CH_4 is largely oxidized to CO_2 within the atmosphere after about 10 years, on average. Carbon is transferred between the atmosphere and the ocean mixed layer through diffusion of gaseous CO_2 across the air–sea interface.

Figure 2.7 illustrates the major processes involved in the transfer of CO_2 between the atmosphere and the deep ocean. When CO_2 enters sea water, the following chemical reactions take place:

$$CO_2 \text{ (gas)} + H_2O \text{ (liquid)} \rightarrow H_2CO_{3(aq)} \text{ (carbonic acid)} \qquad (2.1)$$

$$H_2CO_3 \rightarrow H^+ + HCO_3^- \qquad (2.2)$$

$$CO_3^{2-} + H^+ \rightarrow HCO_3^- \qquad (2.3)$$

giving the net reaction:

$$H_2O + CO_2 + CO_3^{2-} \rightarrow 2HCO_3^- \qquad (2.4)$$

About 90% of oceanic DIC is in the form of HCO_3^-, about 10% is in the form of CO_3^{2-}, and less than 1% is in the form of CO_2. The transfer of CO_2 back to the atmosphere is driven by the partial pressure of CO_2 in the mixed layer (pCO_2), which depends on the concentration of CO_2, not total DIC. The fact that less than 1% of oceanic carbon is in the form of CO_2 means that the pCO_2 in the surface water is much smaller than it would be otherwise. This in turn allows the ocean to hold a large amount of carbon without creating the large back-pressure that would drive CO_2 back into the atmosphere.

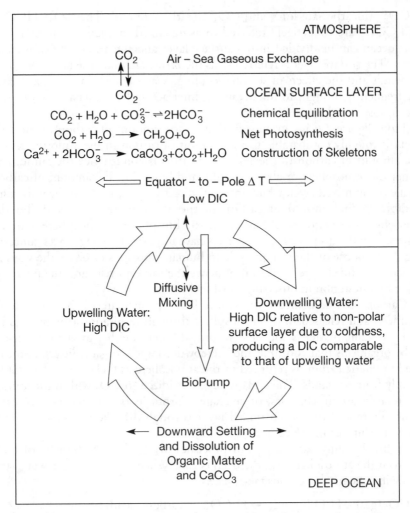

Figure 2.7 Processes that transfer carbon between the atmosphere, oceanic mixed layer, and deep ocean. Reproduced from Harvey (1996a).

In the surface layer of the ocean, two biologically driven processes of importance occur: the construction of soft organic tissues by photosynthesis, which can be represented by the reaction

$$CO_2 + H_2O \rightarrow CH_2O + O_2 \qquad (2.5)$$

and the construction of calcareous skeletons, through the net reaction

$$Ca^{2+} + 2HCO_3{}^- \rightarrow CaCO_3 + H_2O + CO_2 \qquad (2.6)$$

Photosynthesis in the oceans is estimated to take up about 50 Gt C per year. Most of this goes into short-lived micro-organisms, and is rapidly returned to ocean surface water when these organisms die or are eaten and "burned" by larger organisms. However, some of the soft tissue and skeletal material produced in the mixed layer ends up in the deep ocean through sinking dead micro-organisms. The combined flux is crudely estimated to be about 10 Gt C per year, and is referred to as the *biological pump*. About 60–80% of this flux is due to falling soft tissue, and the remainder is due to falling $CaCO_3$ particles. Most of the organic tissue and $CaCO_3$ dissolves in the deep ocean, causing the deep-ocean DIC concentration to be about 10% higher than the DIC concentration in surface waters. As a result, diffusion transfers DIC upward, and this is the main process that balances the downward transfer by the biological pump. Some of the falling $CaCO_3$ survives dissolution and is buried in oceanic sediments – a point we shall return to in Chapter 7 when we discuss how the ocean responds to anthropogenic emissions of CO_2. The biological pump also depletes the mixed layer and enriches the deep ocean in nutrients. The globally averaged variation in both DIC and phosphate (a key nutrient) is shown in Figure 2.8. Both diffusion and advective overturning are important in restoring nutrients to the mixed layer.

We can now see that there are two reasons why the atmospheric CO_2 concentration is so low, and why most of the carbon is in the oceans. The first reason is related to the fact that most of the DIC in the ocean does not occur as dissolved CO_2. The second reason is that the DIC concentration is comparatively low in the surface mixed layer but high in the deeper ocean, combined with the fact that the atmosphere is in direct contact only with the mixed layer but is largely isolated from the deep ocean.

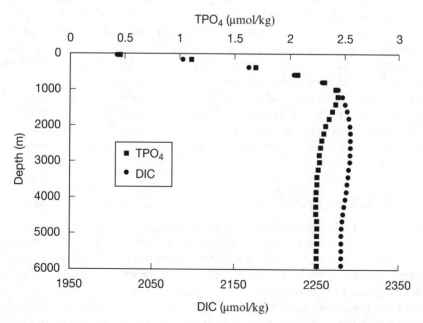

Figure 2.8 The globally averaged vertical profile of dissolved phosphate (from Levitus, 1982) and dissolved inorganic carbon (from Takahashi *et al.*, 1981) in the ocean.

Considerable insight into the behaviour of the carbon cycle can be gained by looking at the ratio of the size of a given reservoir to the total rate of flow of carbon into or out of the reservoir. This ratio is called the *residence time*, τ. As shown in Box 2.2, τ is the average length of time that a C atom spends in a reservoir before being transferred to another reservoir; it is also the length of time required, on average, to completely replace all the C atoms in the reservoir; and it is the length of time required for the amount of carbon in a given reservoir to largely adjust to a change in the carbon flows into or out of the reservoir. From Figure 2.6 it is seen that the amount of carbon in the atmosphere is completely replaced about once every 5 years (τ_{atm} = 5 years) through exchange with either the terrestrial biosphere or mixed layer alone. Conversely, the carbon in the terrestrial biosphere and mixed layer is replaced very rapidly by exchange with the atmosphere ($\tau_{biota} \approx \tau_{ML} \approx$ 6 years), while the average turnover time for soil carbon is slower ($\tau_{soil} \approx$ 20 years). The atmosphere + biota + soils + mixed layer components thus form a tightly coupled subsystem which responds quickly to changes in the fluxes between them. The total amount of carbon in this coupled subsystem is about 3100 Gt. It is linked to the deep ocean through exchanges between the mixed layer and deep ocean – the biological pump and upward diffusion – which amount to about 10 Gt C yr^{-1} in either direction. This gives a turnover time for carbon in the coupled subsystem of about 300 years.

When fossil fuel carbon is added to the atmosphere, the relevant response time scale is the turnover time between the coupled subsystem and the deep ocean (300 years), not the residence time (5 years) of atmospheric carbon based on the exchange with the other reservoirs with which the atmosphere rapidly interacts. This is because the rapid transfer of carbon from the atmosphere to the biota or mixed layer is quickly followed by a return flow to the atmosphere, whereas carbon that enters the deep ocean is effectively removed from the coupled subsystem. This is a very crude representation of highly complex processes, which are discussed in more detail in Chapter 7, but serves to illustrate in an intuitively simple manner how two very different response time scales for atmospheric CO_2 (5–6 years and 300 years) can arise and how we can get a feeling for what the magnitude of the response time scales should be.

2.5 The sulphur cycle

The sulphur cycle involves the emission of sulphur (S) from land plants, the oceans, and volcanoes in many different chemical forms. The main climatic effect of S is through the formation of sulphate (SO_4^{2-}) aerosols, which directly affect climate by reflecting a portion of the incoming solar radiation back to space. They have other, much less certain, effects through their role as cloud condensation nuclei (CCN). The greater the availability of CCN, the easier it is to form cloud droplets. This in turn affects the reflectivity of clouds and their lifespan within the atmosphere.

Box 2.2 Residence time and adjustment time

The ratio of the size of a carbon reservoir to the total carbon flux in or out of the reservoir gives the average turnover time of carbon in the reservoir. Here, we show that, if the reservoir is thoroughly mixed and the outflow varies in direct proportion to the size of the reservoir, then this ratio also represents both the average lifetime of molecules in a reservoir and the time required for the number of molecules originally in the reservoir at time $t = 0$ to decrease to e^{-1} of the number originally present.

If the flux of CO_2 out of a reservoir is linearly proportional to the number of molecules in the reservoir, we can write

$$\frac{dN}{dt} = -kN \tag{2.2.1}$$

where N is the size of the reservoir. For the atmosphere, $N = 770\,Gt$ and dN/dt due to photosynthesis is $110\,Gt\,yr^{-1}$ for present conditions, so $k = 0.14\,yr^{-1}$. It is convenient to introduce a variable τ, given by $1/k$. Then τ (or $1/k$) is none other than the turnover time (flux N divided by throughput dN/dt). Integrating Equation (2.2.1) gives

$$N(t) = N_0 e^{-kt} = N_0 e^{-t/\tau} \tag{2.2.2}$$

which gives the decrease in the number of molecules originally present in the atmosphere at time $t = 0$. When $t = \tau$, N equals $1/e$ times N_0.

The average time $\overline{t}$ spent by a molecule in the reservoir is given by the number of molecules, $N(t)$, in each small range of lifespan dt, summed over all the possible lifespans and divided by the total number of molecules originally present. That is,

$$\overline{t} = \frac{1}{N_0} \int_0^\infty N_0 e^{-t/\tau}\, dt = \tau \tag{2.2.3}$$

In sum, for well-mixed reservoirs where the rate of removal depends on the amount of material present, dividing the steady-state mass of the reservoir by the steady-state outflow gives the amount of time required to completely replace the mass originally present. This is known as the turnover time. The turnover time is equal to the time constant for the exponential decrease in the number of atoms originally present, and also equals the average length of time spent by an atom in the reservoir.

The preceding analysis assumes that the average lifespan τ is constant. However, in the case of CH_4, emissions of the gas into the atmosphere increase the atmospheric lifespan of all the CH_4 already present. As a result, a *perturbation* in the atmospheric concentration of CH_4 decreases more slowly than the exponential decrease given by Equation (2.2.2) using the unperturbed value of τ. This slower decrease is referred to as the *adjustment time*, and is about 20–60% longer than the residence time (Prather *et al.*, 1995, Section 2.10.2.3).

Figure 2.9 shows the components of the sulphur cycle and the associated fluxes that directly involve sulphate. Biological processes on land result in the emission of S primarily as H_2S (hydrogen sulphide), while marine phytoplankton emit S as $S(CH_3)_2$ (dimethylsulphide or DMS). The vast majority of the biogenic S emissions to the atmosphere occur as DMS. Hydrogen sulphide is oxidized by reaction with the hydroxyl radical (OH) to SO_2, while DMS can be oxidized to either methanosulphuric acid (MSA, CH_3SO_3H) or SO_2. About 80% of the SO_2 is oxidized to SO_4^{2-} which, as noted above, forms aerosol particles. The transformation of SO_2 to SO_4^{2-} can occur in the gas phase or inside cloud droplets or sea salt water aerosols, which adds considerable complication to the sulphur cycle. Sulphur is delivered to the ocean in rainwater and in river water as SO_4^{2-}, and occurs in ocean water overwhelmingly as SO_4^{2-}. Oceanic SO_4^{2-} can be directly released to the atmosphere as a component of sea salt aerosols that are formed when air bubbles rise to the ocean surface and burst.

Anthropogenic emissions of S occur largely as SO_2 which, like natural sources of SO_2, is largely converted to sulphate. Sulphate is easily removed from the atmosphere in rainwater, so that the average lifespan of sulphate aerosol in the atmosphere is only about 5 days. This implies, first, that aerosol concentrations (and their climatic effects) will respond essentially instantaneously to changes in

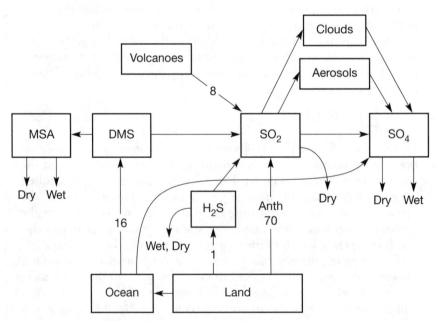

Figure 2.9 Components of the global sulphur cycle involving sulphate (SO_4^{2-}). "Dry" and "Wet" refer to dry and wet deposition, respectively; MSA = methanosulphuric acid; DMS = dimethylsulphide; H_2S = hydrogen sulphide. The numbers next to arrows between boxes are the approximate annual rates of sulphur transfer in Tg S. Based on Schlesinger (1997), Charlson et al. (1992), and other sources.

emissions, and second, that considerable regional variations in the sulphate aerosol concentration are possible, since the atmospheric lifespan is short compared to the length of time required for long-distance transport.

2.6 The nitrogen cycle

Nitrogen is an essential element for life, and it is important to climate through its effect on the rate of photosynthesis on land and in the ocean, through the greenhouse effect of N_2O (nitrous oxide), through its role in the chemistry of ozone in both the stratosphere and troposphere, and as a likely source of CCN.

The nitrogen cycle and the major processes involved in it are illustrated in Figure 2.10. The atmosphere is 78% N_2, but nitrogen in this form cannot be used by plants. Rather, nitrogen must first be converted to NH_4^+ (ammonium) by a process called *nitrogen fixation*, which is carried out by several types of bacteria and by blue-green algae. Some of these nitrogen fixers exist freely in soils (asymbiotic), while others form symbiotic relationships with the roots of higher plants. Once converted to NH_4^+, nitrogen can be taken up by plants through direct absorption of ammonium, or through *nitrification* – in which bacteria convert NH_4^+ to nitrite (NO_2) and nitrate (NO_3^-) – followed by absorption of nitrate and its assimilation into organic matter. During the decomposition of organic matter, *mineralization*, or the release of nutrients in inorganic forms, occurs. This is largely carried out by microbes, which are very efficient in absorbing the nutrients that are released by mineralization,

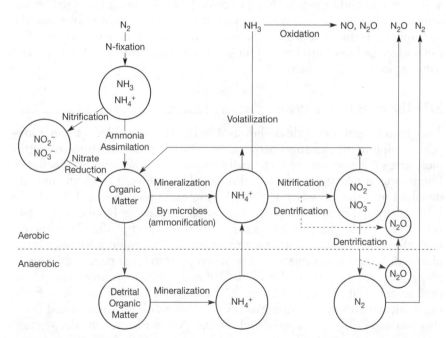

Figure 2.10 Biological processes involved in the global nitrogen cycle, based on Schlesinger (1991, Figure 12.1) and Jaffe (1992, Figure 12.1).

resulting in *immobilization* of the nutrients. Some of the NH_4^+ that is released by mineralization and not taken up by plants or microbes is eventually converted to N_2O (nitrous oxide). N_2O can diffuse to the stratosphere, where it catalyses the destruction of O_3 (another GHG, important for shielding life from harmful ultraviolet radiation) and is destroyed by ultraviolet radiation.

Nitrogen fixation in the sea is less intense than on land, amounting to about 40 Tg N/yr globally (compared to about 150 Tg N/yr on land). N_2O is produced in the ocean in much the same way that it is produced in soils, and diffuses to the atmosphere in association with both nitrification and denitrification. In most regions of the ocean, nitrate is not measurable in the surface water, suggesting that nitrogen is a limiting nutrient for biological productivity in the oceans (Schlesinger, 1997).

Humans have perturbed the N cycle through direct emissions of NO_x (the sum of NO and NO_2) and N_2O, and through the production of nitrogen fertilizers. Natural nitrogen fixation on land is estimated to be about 150 Tg N per year, with another 40 Tg N per year due to the production of fertilizers. Some of the fixed nitrogen is returned to the atmosphere as N_2O through the process of denitrification, so the production of fertilizers has increased the rate of production of N_2O. Industrial processes and biomass burning also create N_2O (a more detailed breakdown of anthropogenic emissions of N_2O is given in Section 3.1). As noted above, N_2O is destroyed when it reaches the stratosphere. The combustion of fossil fuels also adds about 40 Tg N per year to the atmosphere as NO_x, much of which is redeposited on land as a component of acid rain. This anthropogenic N input could have stimulated terrestrial photosynthesis enough to remove an extra 0.5–1.0 Gt C per year from the atmosphere, as discussed later in Section 7.4. However, excessive inputs of N can lead to the loss of mycorrhizal fungi, thereby exacerbating P deficiency and reducing long-term productivity.

2.7 The phosphorus cycle

The global phosphorus cycle is illustrated in Figure 2.11. Unlike carbon, nitrogen, or sulphur, the exchanges between the atmosphere and either the terrestrial biosphere or the oceans are very small compared to other phosphorus fluxes. Phosphorus is released from rocks through chemical weathering on land and dissolved into the soil water as dissolved inorganic phosphorus (DIP). Some DIP is delivered to the oceans by rivers, some is taken up by plants and converted to organic phosphorus (that is, it becomes part of the plant organic matter), and the balance reacts with other minerals to produce forms of phosphorus that are not available to plants. When plant material decays, the organic phosphorus can be converted back to DIP and reused by plants. Some of the dead organic matter is carried by rivers to the oceans as small particles – forming particulate organic phosphorus (POP). Marine photosynthesis and decay also produce POP. As terrestrial and marine POP settle into the deep ocean, some will be consumed along the way, while the rest will be buried and eventually become part of new rocks, thereby completing the cycle.

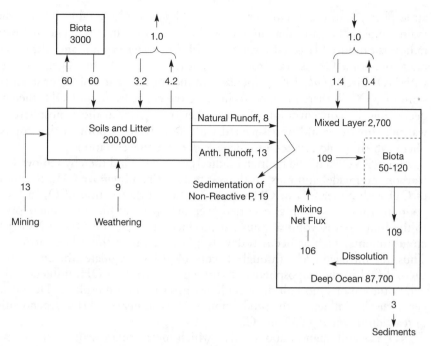

Figure 2.11 The global phosphorus cycle. Numbers in boxes are the amounts of phosphorus in each reservoir as Tg P, while numbers next to arrows between boxes are the annual rates of phosphorus transfer in Tg P. Based on Schlesinger (1991, 1997), with some fluxes adjusted so that the natural cycle is in steady state.

Both nitrogen and phosphorus appear to be limiting nutrients for marine biological productivity. Phosphorus is also likely to be a limiting nutrient for biological productivity in tropical terrestrial ecosystems. The residence time of nitrogen in the ocean is on the order of years, whereas that of phosphorus is on the order of 30,000 years. However, rapid changes in the surface concentration of either nutrient can occur through changes in the rate of upwelling. This in turn could lead to important changes in the strength of the biological carbon pump and hence in the amount of CO_2 in the atmosphere.

2.8 Linkages between the biogeochemical cycles through atmospheric chemistry

Chemical reactions in the atmosphere provide a number of linkages between the carbon, nitrogen, and sulphur cycles. Some of the more important linkages will be discussed here in order to illustrate the importance of atmospheric chemistry in the changing composition of the atmosphere.

A central player in tropospheric chemistry is the hydroxyl radical (OH), which has a concentration of only 3×10^{-14} but is the main oxidizing agent in the atmosphere. It oxidizes and thereby destroys most trace gases in the atmosphere, with the notable exception of CO_2 and N_2O. Table 2.1 lists the main atmospheric gases that are removed by reaction with OH. The hydroxyl radical is produced largely by reaction with O_3, water vapour, and ultraviolet radiation. Some OH is also produced by reaction of HO_2 (hydrogen dioxide) with NO (Eisele et al., 1997). Increases in the concentration of O_3 or in emissions of NO will therefore tend to increase the concentration of OH, thereby decreasing the concentration of most other trace gases in the atmosphere. A warmer climate would be associated with more water vapour in the troposphere, which would also tend to increase the OH concentration.

About 50–60% of the hydroxyl radicals react with and thereby remove CO (carbon monoxide) and most of the rest react with and remove CH_4. Reaction of OH with either CO or CH_4 results in the net destruction of O_3 in NO-poor environments, such as occurred over large parts of the troposphere prior to the Industrial Revolution, but net production of O_3 and OH in NO-rich environments, such as occur today over large areas of the NH continents. Thus, the nitrogen cycle, through its control of NO, regulates the amount of O_3 and OH in the troposphere and, through its effect on OH, influences the concentration of almost all the greenhouse gases in the atmosphere. The nitrogen cycle also influences the production of O_3 (and hence OH) independently of reactions involving CO and CH_4.

NO_x has a lifespan of about 1 day, which means that essentially none can reach the stratosphere. However, N_2O can reach the stratosphere (its lifespan is 120 years), where it is dissociated by ultraviolet radiation into NO through the reactions.

Table 2.1 Selected atmospheric trace gases which are removed in part or entirely through reaction with OH, from Schlesinger (1991), except where indicated. NMHCs = non-methane hydrocarbons

Trace gas	Concentration in NH (ppbv)	Tropospheric lifespan	Contribution of OH-sink reaction to removal (%)
CH_4	1780	7.9 years[a]	90
CO	250	20–50 days[b]	100
NMHCs[e]	2–10	1–100 days	50–100
SO_2	0.2	2.4 days[c]	20[c]
COS	0.5	5 years	30
H_2S	—	4 days	100
$(CH_3)_2S$	—	1 day	50
NO, NO_2	0.1	1 day	100[d]
NH_3	1	14 days	10

[a] Lelieveld et al. (1998).
[b] Mauzerall et al. (1998).
[c] Roelofs et al. (1998).
[d] According to Wang et al. (1998b), half of NO_x is removed by hydrolysis of N_2O_5 in aerosols and half by reaction with OH.
[e] NMHCs = Non-methane hydrocarbons.

$$N_2O + h\nu \rightarrow N_2 + O \tag{2.7}$$

and

$$N_2O + O \rightarrow 2NO \tag{2.8}$$

NO then reacts with and destroys O_3 through the reaction

$$NO + O_3 \rightarrow NO_2 + O_2 \tag{2.9}$$

This converts NO to NO_2, which then forms HNO_3 and diffuses to the troposphere, where it is washed out in rain. Photochemical dissociation of N_2O in the stratosphere is the only removal process for N_2O (except for a possible ocean sink), which explains its long atmospheric lifespan. From the above, it can be seen that N emissions (as NO_x) lead to higher O_3 concentrations in the troposphere, while N emissions (as N_2O) lead to the destruction of O_3 in the stratosphere.

2.9 Linkages between the biogeochemical cycles and climate

The preceding sections outlined the flows of energy and mass (as H_2O, C, S, N, and P) that link the various components of the climate system. The various energy and mass flows are closely interlinked through a variety of *feedback* processes, and in this section some of the major linkages are identified.

A feedback occurs when a change in some quantity, ΔA, leads to a change in a second quantity, ΔB, and the change ΔB tends to provoke a further change in A. If ΔB provokes a ΔA that is in the same direction as the initial change, then a *positive feedback* occurs, whereas if it provokes a ΔA that is opposite to the initial change, a *negative feedback* occurs. Positive feedbacks amplify the initial change and tend to destabilize a system, whereas negative feedbacks dampen the initial change, thereby stabilizing a system.

The tightest linkage is between the energy and water vapour flows, in that evaporation and condensation of water are the major processes that offset the radiative heating of the surface and the radiative cooling of the atmosphere. As the ocean surface temperature increases, relatively more of the net radiation goes into evaporating water, so that the overall intensity of the hydrological cycle and the atmospheric water vapour content tend to increase. Since water vapour is a greenhouse gas, this leads to further warming, a further increase in evaporation, and so on, as part of a positive feedback loop.

Other known or potential feedback loops involving climate and the biogeochemical cycles are the following:

- The rates of photosynthesis and respiration on land are directly affected by climate, through their dependence on both temperature and soil moisture, and through the dependence of photosynthesis on cloud cover. An increase in photosynthesis will tend to reduce the atmospheric CO_2 concentration, while an increase in the rate of respiration will tend to increase atmospheric CO_2, thereby provoking further changes in climate.

- Climate is likely to affect the strength of the oceanic biological pump through its effect on the rate of vertical overturning circulation and turbulent mixing within the oceans, which supply the nutrients to the surface layer that are needed for photosynthesis.
- The atmospheric methane concentration depends on the magnitude of the emission sources and on the rate of removal. A warmer climate could increase CH_4 emissions from wetlands by increasing the rate of photosynthesis and hence the supply of organic matter for decomposition, and through the direct effect of warmer temperatures on the rate of decomposition. A warmer climate would also increase the rate of removal of CH_4 from the atmosphere, by increasing the OH concentration (in turn a result of a higher water vapour content in the atmosphere).
- Changes in climate or in surface-water nutrient concentrations (due to a climate-induced change in the rate of upwelling) could affect the production of DMS by marine bacteria, thereby altering the concentration of sulphate CCN and changing the optical properties of clouds. However, the direction or magnitude of these possible effects is unknown.

2.10 Natural causes of climatic change

Climate can change naturally for a variety of reasons. Some of the driving factors of climatic change operate at time scales of hundreds of millions of years, whereas others fluctuate over a time period of only a few years. The main causes of natural climatic change are as follows.

Changes in the composition of the Earth's atmosphere

The atmospheric CO_2 content can change as a result of driving forces that are external to the climatic system, or as a result of internal feedbacks between climate and the carbon cycle. Externally driven changes in atmospheric CO_2 include (i) long-term changes in the rate of volcanic degassing of CO_2; (ii) an increase in the rate of weathering on land (which consumes CO_2 as part of the chemical reactions that occur during weathering), in turn a result of tectonically driven periods of mountain building; and (iii) a change in the strength of the oceanic biological pump due to a change in the latitudinal and seasonal distribution of solar radiation as the Earth's orbit changes. An example of an internally driven change would be a change in the strength of the biological pump due to some internally generated change in the rate of upwelling of nutrient-rich deep water.

Changes in topography, land–sea geography, and bathymetry

The topography of the land surface affects wind and rainfall patterns, thereby exerting an immediate effect on weather patterns and climate, and – at high latitudes – determines the ease or difficulty of initiating ice ages. Continental drift leads to changes in the proportion of ocean and land at different latitudes. Since continents tend to absorb less solar radiation than oceans, shifting a continent from low latitudes to higher latitudes (where there is less radiation to be reflected) results in a net increase in the global average absorption of solar radiation, thereby tending to warm the global mean climate. Changes in the shape of ocean basins (bathymetry) and, in particular, such things as the closing of the former gap between North and South America or the opening of the gap between Antarctica and South America, altered ocean circulation patterns worldwide. The altered circulation patterns in turn led to large changes in regional climate by redistributing heat from one region to another. In cases where large new areas of snowcover developed (such as with cooling of Antarctica), this would have also led to changes in the globally averaged climate.

Changes in solar luminosity

The energy output from the sun varies on at least two different time scales. At one extreme, the luminosity has systematically increased from about 70–75% of its present value 3.5 billion years ago, to its present value (Newman and Rood, 1977; Gough, 1981). At the other extreme, the solar luminosity has been observed to vary by about ±0.07% over the 11-year sunspot cycle (Hoffert et al., 1988). Observations of other stars suggest that variations of a few tenths of a percent could occur in the sun over an 80-year period (Baliunas and Jastrow, 1990). Changes in the flux of ultraviolet radiation, in association with overall changes in solar luminosity, might induce changes in the amount of stratospheric O_3 that significantly reduce the net heating or cooling effect of changes in solar luminosity (Haigh, 1994).

Changes in the Earth's orbit

Owing to the perturbing influence of the gravitational force from other planets in the solar system, the Earth's orbit changes in a systematic and calculable way. Changes in the Earth's orbit are regarded as the driving mechanism behind glacial–interglacial oscillations in climate, although several feedback mechanisms – including changes in the atmospheric concentrations of CO_2 and CH_4 – are required to fully explain ice age cooling.

Volcanic activity

Volcanic activity influences climate through the emission of SO_2 during eruptions. The SO_2 is converted to sulphate aerosols once in the atmosphere, but because much of the SO_2 is injected into the stratosphere (rather than the troposphere), the aerosols persist for a few years rather than a few days and are therefore able to spread globally. These aerosols cool the climate by reflecting sunlight, but the cooling effect – like the aerosols themselves – lasts only a few years. Major volcanic eruptions during the past 140 years caused a global mean cooling of 0.1–0.2°C during the first two years after the eruption, although North America and Eurasia warmed by several degrees in the first or second winter after major eruptions due to shifts in wind patterns (Robock and Mao, 1995).

Internal variability of the atmosphere–ocean system

A number of computer models of the atmosphere–ocean system have been built in which self-sustained oscillations in climate occur without any external driving force. Oscillations at a time scale of several hundred to several thousand years involve alternating heat flows into or out of the deep ocean. Oscillations at longer time scales can be generated through interactions between the oceans and continental-scale ice sheets, while decadal-scale fluctuations in regional climate can arise through interactions between the atmosphere, ocean surface temperature, and the extent of sea ice (Mysak and Manak, 1989).

Internally generated climatic variability is largest at the subcontinental scale, but decreases in magnitude as the scale under consideration increases. Thus, year-to-year fluctuations in regional temperature of a few degrees are common, but year-to-year fluctuations in global mean temperature are generally less than 0.1 or 0.2°C. This is because much of the internally generated variability involves redistributing heat horizontally from one region to another, effects which cancel out when averaged over the entire planet.

2.11 Human causes of climatic change

Humans are altering climate at a global scale by altering the atmospheric concentrations of GHGs, both by direct emissions of GHGs into the atmosphere, and indirectly by inducing changes in the chemistry of the atmosphere which lead to changes in the concentrations of some GHGs. Humans also affect the climate on a continental scale through the emission of a variety of aerosols. Changes in land cover (e.g., deforestation) also influence climate, largely at the local scale but possibly with some global-scale effects through large-scale waves in the atmosphere.

Water vapour is the strongest contributor to the natural greenhouse effect, but, among the GHGs, it is most directly linked to climate and therefore least directly controlled by human activity. This is because evaporation is strongly dependent on surface temperature, and because water vapour cycles through the atmosphere quite rapidly, once every eight days on average. Concentrations of the other GHGs are strongly and directly influenced by emissions associated with the combustion of fossil fuels, by some agricultural activities, and by the production and use of various chemicals.

Increasing concentrations of well-mixed GHGs

Those GHGs that remain in the atmosphere long enough (one year or more) that they can be fairly thoroughly mixed by atmospheric winds before being removed have a uniform concentration (to within a few percent) within the atmosphere. Such gases are said to be *well mixed*. All of the greenhouse gases except O_3 and a few minor gases have atmospheric lifespans well in excess of one year, and so are well mixed. A few greenhouse gases (such as chloroform, HCFC-123, and HFC-152a) have a lifespan of around one year; these gases are fairly well mixed within the troposphere, but have a sharply lower concentration within the stratosphere.

The lifespan of a gas plays an important role in determining its effect on climate, since the greater the lifespan, the greater the buildup of the gas in the atmosphere that will occur for a given rate of emission. The relationship between emission rate, lifespan, and concentration is explored in Box 2.3. The other factor determining the climatic effect of a given gas is its ability to trap heat on a molecule-by-molecule basis. Table 2.2 lists the important well-mixed GHGs, the average atmospheric lifespan for each gas, and the ability of each gas to trap heat on a molecule-per-molecule basis compared to the heat-trapping ability of CO_2.[1] Many are several hundred to several thousand times more effective in trapping heat than is CO_2, and some of the GHGs will remain for hundreds to thousands of years in the atmosphere. The last three

---

Box 2.3 Relationship between lifespan and steady–state concentration

The atmospheric lifespan of most greenhouse gases is long enough that the atmosphere can be treated as a well-mixed reservoir, and the rate of removal of the gas varies directly with the concentration. Changes in the concentration, C, of the gas in the atmosphere are given by the imbalance between the flux input, F, and the rate of removal, C/τ. That is,

$$\frac{dC}{dt} = F - \frac{C}{\tau} \qquad (2.3.1)$$

where τ is the average residence time (or turnover time). The steady-state concentration, C_0, is such that $dC/dt = 0$, from which it follows that

$$C_0 = \tau F_0 \qquad (2.3.2)$$

where F_0 is the initial flux input. If there is a perturbation in the flux, such that $F = F_0 + \Delta F$, then the new steady-state concentration will be

$$C' = (C + \Delta C) = \tau(F + \Delta F) \qquad (2.3.3)$$

Subtracting Equation (2.3.2) from Equation (2.3.3) gives the result that $\Delta C = \Delta F$.

Thus, for a given additional flux into the atmosphere, the steady-state change in the concentration of the gas varies directly with the average lifespan of the gas. This is because the longer the lifespan, the greater the buildup in gas concentration that is required before the rate of removal (which is proportional to the amount of gas present) equals the rate of input.

---

1   As previously noted, the removal processes for CO_2 are considerably more complicated than for other gases, and it is not correct to speak of a single average lifetime in the case of CO_2. This issue and the specific removal processes for other gases are discussed in Chapter 7.

Table 2.2 Greenhouse gases directly emitted through human activities, relative heat trapping on a molecule-per-molecule basis compared to CO_2, average lifetime in the atmosphere, concentration in 1995, and global mean radiative heating in 1995

Gas	Relative heat trapping ability[a]	Atmospheric lifespan (years)	Concentration (ppbv) Pre-industrial	1995	Heating perturbation in 1995 ($W\,m^{-2}$)
CO_2	1	Variable	278,000	360,000	1.40
CH_4	26	7.9[b]	700	1725	0.47
N_2O	206	120	275	311	0.14
CFC-11	12,400	50	0.000	0.272	0.082
CFC-12	15,800	102	0.000	0.532	0.205
HFC134a	9570	14.6	0.000	0.0016	0.0003
SF_6	≤36,000	3200	0.000	0.0032	0.0020
CF_4	5600	50,000	0.000	0.075	0.0071

[a] From Shine *et al.* (1990), except for CH_4 (Lelieveld and Crutzen, 1992), SF_6 (derived from Ko *et al.*, (1993), and CF_4 (derived from Schimel *et al.*, 1996).

[b] Based on Lelieveld *et al.* (1998). The value of 12.2 years given in Schimel *et al.* (1996) is the adjustment time. The difference between the atmospheric lifespan and the adjustment time is explained in Box 2.2.

columns of Table 2.2 give the pre-industrial concentration, the concentration in 1995, and the estimated globally averaged rate of heat trapping in 1995. The 1995 heat trapping depends on the effectiveness of the gas in trapping heat and the increase in concentration that occurred up to 1995; the latter depends on the cumulative emission, the timing of past emissions, and the average lifespan of the gas in the atmosphere. Although CO_2 is the least effective in trapping heat on a molecule-per-molecule basis, the amount of CO_2 emitted is so much larger than that of any other GHGs that the CO_2 increase accounts for about 55% of the total heat trapping that has occurred so far.

Changes in the concentration of ozone

Ozone differs from the other greenhouse gases in that its atmospheric lifespan is relatively short – from 10 days near the surface in the tropics, to 200 days in polar regions and at high altitude. As a result, O_3 cannot spread very far before it is removed, so large differences in its concentration occur from one region to another. Ozone also differs from the other GHGs in that it is not directly emitted into the atmosphere; rather, it is produced through photochemical reactions involving other substances – referred to as *precursors* – that are directly emitted. There are also several chemical reactions involved in the destruction of O_3, and the O_3 concentration at any given place and time depends on the balance between production, destruction, and transport to or from other regions. The effectiveness of the gases involved in producing or destroying ozone depends on where they occur and on the concentrations of other gases involved in the chemistry of O_3.

Increasing concentration of aerosols

Aerosols, like GHGs, are produced both naturally and through human activity; natural aerosols include sea salt, dust, and volcanic aerosols, while anthropogenic aerosols are produced from the burning of biomass and fossil fuels, among other sources. Some aerosols, such as dust, are directly emitted into the atmosphere. The majority of aerosols, however, are not directly emitted but – like tropospheric O_3 – are produced through chemical transformation of precursor gases. The most important anthropogenic aerosol is thought to be sulphate, which is produced from SO_2 that is released with the combustion of sulphur-containing coal, from the refining of oil, and from the smelting of certain metals. Other anthropogenic aerosols include non-absorbing organic compounds produced from the burning of biomass and the oxidation of hydrocarbons, light-absorbing aerosols (soot or black carbon) produced from the incomplete combustion of fossil fuels and biomass, nitrate aerosols, and anthropogenically induced increases in the production of dust. All tropospheric aerosols have a short lifespan (days) in the atmosphere owing to the fact that they are rapidly washed out with rain. For this reason and because the strength of the emission sources varies strongly from one region to another, the amount of aerosols in the atmosphere varies considerably from one region to another.

Changes in the land surface

It has long been suggested that human-induced changes in the land surface albedo (reflectivity) could have had a noticeable effect on regional- and global-scale temperature. Among the first estimates of such effects are those of Sagan *et al.* (1979), who suggested that changes in land cover (primarily desertification, deforestation, and salinization) caused a global mean cooling tendency of 0.2°C during the preceding 25 years, and up to 1°C cooling during the preceding several thousand years. A more recent analysis by Bonan (1997), focusing on the USA, indicates that conversion of forest to cropland caused a summer cooling of up to 2°C over a wide region of the central United States. Deforestation in the tropics may also have had significant climatic effects, both locally and in NH mid-latitudes over land through induced changes in atmospheric winds (Chase *et al.*, 1999).

2.12 Comparison of natural and future human causes of climatic change

Natural and human causes of climatic change can be compared in terms of the magnitude of the radiative heating perturbation and the associated climatic change, and in terms of the time scale over which such changes occurred. In Table 2.2 the heat trapping due to the buildup of GHGs that has already occurred was listed. This heat trapping, which amounted to about $2.4\,W\,m^{-2}$ by 1990, could eventually reach $5–10\,W\,m^{-2}$ and produce a globally averaged warming of 2–10°C during the next two centuries. This is not likely to be offset to any significant extent by the cooling effect of sulphate aerosols, since concerns over acid rain will almost surely strongly limit sulphur emissions in the future. Table 2.3 summarizes the various natural causes of climatic change in terms of the time scales and magnitude of change. It can be seen that natural climatic changes comparable to or greater in magnitude than that expected over the next 200 years have occurred. However, natural climatic changes have generally occurred over much longer periods of time (typically requiring hundreds of thousands to millions of years). The only exception is for relatively rare abrupt changes, which, were they to occur during the 21st century, would be catastrophic. Thus, although natural processes might eventually lead to a new ice age, this and other natural tendencies will be swamped by the warming effect of increasing concentrations of GHGs during the next few centuries.

Table 2.3 Summary of the major natural causes of climatic change in terms of time scale and magnitude

Cause	Time scale (years)	Global mean magnitude, lower and upper limits
Volcanic activity	1–4	0.4°C[a]
Internal variability	10^1–10^3	0.2–0.4°C[b]
Transitions to new climate states[c]	10^1–10^2	2–3°C[d]
Changes in solar luminosity	10^1–10^9	0.1°C[e]
Changes in GHG concentrations	10^2–10^9	2–3°C[f]
Changes in the Earth's orbit	10^4–10^5	4–6°C[g]
Changes in land–sea geography	10^7–10^8	up to 5°C[h]

[a] This is the largest inferred effect during the past 130 years.

[b] Based on millennial-scale simulations with computer climate models and multi-century reconstructions of NH mean surface-air temperature using proxy paleoclimatic data.

[c] The time scale here is the fastest time seen for the transition from one climatic state to another in the paleoclimatic record. The timing of the transitions is highly variable, and such transitions might be triggered by slower and more gradual climatic changes.

[d] Based on the finding of transitions in polar ice cores corresponding to roughly half of the difference between glacial and interglacial climates in these regions. If the cause of the polar climatic change is local (e.g., from a change in nearby ocean current), then the global mean temperature change as a fraction of the global mean difference between glacial and interglacial climates would have been smaller.

[e] This is a reasonable estimate of the effect of solar variability of no more than a few tenths of a percent during the past 100 years (see Section 9.2). Since solar luminosity is estimated to have increased from 70% of its present value to its present value of the past 3 billion years, the associated change in temperature would have been extremely large were it not for largely counteracting changes in other variables (such as atmospheric GHG concentrations).

[f] Based on estimates that decreases in GHG concentrations during the peak of the last ice age can explain about half of the inferred global mean cooling of 4–6°C. Much larger changes in CO_2 concentration (up to 4–10 times the present concentration) occurred over the past 500 million years, as illustrated in Fig. 1.2.

[g] This is the estimated difference in globally averaged surface temperature between glacial and interglacial periods. The glacial–interglacial transitions were triggered by changes in the Earth's orbit, but a number of slow positive feedback mechanisms – including changes in GHG concentrations and expansion of ice sheets – are required to explain the full amplitude of temperature change.

[h] Based on the maximum simulated effect of changes in the distribution of continents between now and the Cretaceous Period, when the globally averaged surface temperature was at least 6°C warmer than today.

Questions for further thought

1. Explain how local regions could experience cooling as the overall global climate warms up.
2. What are the main differences between the hydrological and carbon cycles?
3. In what ways are the cycles of CO_2 and CH_4 linked?
4. How does the sulphur cycle affect clouds, and how do clouds affect the sulphur cycle?
5. What are some of the distinctive features of (a) the nitrogen cycle, and (b) the phosphorous cycle, compared to the other biogeochemical cycles?
6. How does climate affect atmospheric OH, and how does atmospheric OH affect climate?
7. Why are volcanic eruptions considered to be an *external* cause of climatic change?

Chapter 3

Factors driving anthropogenic emissions to the atmosphere

In the preceding chapter we identified the major natural and anthropogenic causes of climatic change. In this chapter, more information is presented concerning the emissions of GHGs and of ozone and aerosol precursors as a result of human activities. This is followed by data on the historical variation of CO_2 emissions and CO_2 concentrations, which, combined with other evidence to be discussed here, establishes beyond doubt that the observed buildup of CO_2 is due to human emissions of CO_2. The variation in the atmospheric CH_4 and N_2O concentrations since the Industrial Revolution and estimates of the variation in the emissions of aerosol precursors are also presented.

3.1 Emission sources and emission inventories for GHGs, ozone precursors, and aerosols or aerosol arecursors

In this section we identify the major human activities that contribute to emissions of GHGs, ozone precursors, and sulphate aerosol precursors; and we present emission factors for selected energy sources or industrial activities.

Carbon dioxide

When fossil fuels are burned to produce energy, the carbon in the fuel combines with oxygen in the air to produce carbon dioxide. The production of CO_2 is thus inextricably tied with the very use of fossil fuels, and the only way to produce less CO_2 from a given fossil fuel is to use less of it. The three major types of fossil fuels are coal, oil, and natural gas.

Table 3.1 gives the CO_2 emission per unit of energy for various fossil fuels (and in the production of certain chemical products). Among the fossil fuels, coal produces the most CO_2 per unit of energy provided, while natural gas produces the least. Renewable energy sources such as solar and wind energy entail no direct emissions of CO_2. Conversely, the production and use of synthetic oil and gas made from coal entails substantially greater CO_2 emissions than the use of conventional oil and gas or the direct use of coal.

A second major source of CO_2 emissions is deforestation. The emissions arise in this case either through the decay of plant debris left on the ground after the extraction of timber, or as a result of forest fires that are set in order to clear land for agricultural purposes (a common practice in tropical countries).

Table 3.1 Carbon dioxide emission factors for combustion of various fossil fuels (given as kg C per GJ of fuel energy) and the manufacture of various chemical products (given as kg C per kg of manufactured product). The emission factors in the latter case reflect CO_2 released during the chemical reactions involved in the manufacture of the product, but do not include emissions associated with the energy needed to make the product. Combustion emission factors are based on the higher heating value of the fuel. Source: see references below

Source	CO_2 emission factor
Combustion sources (kg C/GJ energy)	
Anthracite coal	23.5–26.6[a,b]
Lignite coal	22.2–25.9[c,d]
Bituminous coal	23.9–24.5[d]
Sub-bituminous coal	24.8–25.7[c,d]
Oil	17–20[a,b]
Natural gas	13.5-14.0[a,b]
Process sources (tonne C/tonne product)	
Cement	0.08-0.26[a,e]
Lime	0.215[a]
Ammonia	0.431[a]

[a] Jaques (1992).
[b] Nakicenovic *et al.* (1996, p. 75).
[c] Calculations based on information given by DOE (1992, p. 124).
[d] Calculations based on information given by IEA (1991a, p. 24).
[e] Chen (1998).

The resulting emissions are quite uncertain. A third major source of CO_2 emissions is the chemical release of CO_2 in the production of cement, lime, and ammonia. This CO_2 is in addition to the CO_2 released from the combustion of the fossil fuels that are used to provide the energy for the manufacture of these products.

Table 3.2 presents an inventory of CO_2 emissions from fossil use, chemical products, and land use changes for the 20 countries with the largest total emissions in 1990. The emission estimates were obtained from World Resources Institute (1998), except where indicated. Annual per capita CO_2 emissions range from a low of 1.2 tonnes CO_2 per year (India) to a high of 23.5 tonnes CO_2 per year (Australia). Table 3.3 gives the total global emissions of CO_2 (by mass of carbon)[1] due to each type of fossil fuel, due to the manufacture of cement, and due to land use changes. The global fossil fuel emission of CO_2 is currently about 6.0 Gt C per year, while tropical deforestation is estimated to result in emissions of 1.6 ± 1.0 Gt C per year. Given a current human population of about 6 billion, the world average fossil fuel emission is about one tonne of carbon per person per year (or 3.67 tonnes of CO_2 per person per year).

[1] To convert from tonnes of C to tonnes of CO_2, multiply by 3.67.

Table 3.2 National inventories of CO_2 emission (Mt CO_2) associated with use of fossil fuels; manufacture of cement, ammonia, and lime; and deforestation. Inventories are given for the 20 countries with the largest total emissions. Also given are estimated total CH_4 emissions (Mt CH_4). Source: WRI (1998), except where indicated

| Country | CO_2 emission (Mt CO_2) | | | | Per capita CO_2 (t CO_2) | Methane emission (Mt CH_4) |
	Fossil fuels	Chemical products	Land use changes	Total		
Australia	286.8	3.0	130	420	23.5	6.24
Brazil	236.5	12.7	1200[a]	1450	9.3	—
Canada	430.4	5.3	—	436	14.8	3.51
China	2970.4	222.0	9	3200	2.7	33.83
France	329.6	10.5	−37	303	5.3	2.83
Germany	815.2	19.9	−20	815	10.0	5.20
India	873.9	34.9	150[a]	1060	1.2	—
Indonesia	286.4	9.7	455[a,b]	750	3.8	3.75
Iran	255.7	8.1	—	264	3.8	—
Italy	392.5	17.4	−37	373	6.6	3.91
Japan	1081.7	45.1	−90	1037	8.3	1.32
Mexico	345.9	11.9	89	447	4.9	2.98
North Korea	255.7	1.3	—	257	11.1	—
Poland	331.1	6.9	—	338	8.8	2.47
Russia	1799.9	18.1	−587	1230	8.3	27.36
South Korea	346.1	1.3	—	345	8.3	—
South Africa	301.3	4.5	—	306	7.4	—
Ukraine	432.7	5.5	52	490	7.5	9.46
United Kingdom	535.9	6.2	−6	536	9.2	3.88
United States	5430.2	38.3	−532	4936	18.5	28.17

[a] Taken from Myers (1989).
[b] WRI has Indonesia an implausibly large sink of 822 Mt CO_2, when the country is almost certainly a large source.

Table 3.3 Total global CO_2 emissions (Gt C) in 1995 as a result of the combustion of solid, liquid, and gaseous fossil fuels, as a result of gas flaring and cement manufacture, and due to land use changes. Taken from Marland *et al.* (1998), except for the land use emission, which is taken from Schimel *et al.* (1996)

Source	Emission (Gt C)
Solids	2.45
Liquids	2.57
Gases	1.14
Gas flaring	0.06
Total fossil fuel	6.22
Cement manufacture	0.19
Land use changes	0.6–2.6
Total	7.0–9.0

Methane

Methane is emitted to the atmosphere as a byproduct of fossil fuel use and through a variety of other human activities. Table 3.4 lists methane emission factors per unit of energy extracted from the ground for typical present-day conditions and practices. Methane is produced whenever organic matter decomposes in the absence of oxygen (anaerobic conditions). Coal consists of former swamp deposits that were buried and underwent partial decomposition, so that methane is trapped in most coal seams. The process of mining coal allows this methane to seep into the atmosphere. Methane is found in association with oil, and is often flared during the extraction of oil. Natural gas is 90–95% methane, and some methane leaks into the atmosphere during the extraction and distribution of natural gas. Methane is also released from the slow decomposition of organic matter in land that is flooded for hydro-electric power production. Conservative estimates indicate that the global warming effect of hydro-electric dams in the Brazilian Amazon can be *several times* the global warming effect of fossil fuel power plants producing the same amount of electricity (Fearnside, 1995; Rosa and Schaeffer, 1995).

Methane is released to the atmosphere from a wide range of human activities other than the production of energy. These include the burning of biomass and a number of activities where partial decomposition of organic matter under anaerobic conditions occurs. The latter includes cultivation of rice, the raising of ruminant animals such as cattle, and the operation of landfills and sewage treatment plants. Estimates of the global inventories of methane emission due to various human activities, as summarized by Prather *et al.* (1995), are given in Table 3.5. In most cases the uncertainties are a factor of 3–4 or more, with an overall uncertainty of a factor of 3. The last column of Table 3.2 gives estimated methane emissions by 20 of the countries with the largest emissions.

Table 3.4 Methane emission factors for various fossil fuels .
Source: see references below.

Fuel	CH_4 emission factor (kg CH_4/GJ)
Coal mining	0.13–0.53[a]
Underground	0.46–0.49[b]
Surface	0.12–0.13[b]
Oil	≤0.03[c,d]
Natural gas	0.18–0.19[c,e,f]

[a] Based on CH_4 emission factors given in Beck *et al.* (1993) and references therein.

[b] Based on 1989 coal production with CH_4 emission estimates from Kirchgessner *et al.* (1993). The emission factor depends on the average annual coal heating value assumed.

[c] Barnes and Edmonds (1990).

[d] Emissions occur when gas is vented but not flared, during oil extraction.

[e] Gas Research Institute (1997).

[f] This is the emission factor for each 1% leakage of natural gas, from the point of extraction to the point of use.

Table 3.5 Estimates of global anthropogenic methane emissions from various sources, as summarized by Prather *et al.* (1995).

Methane source	Emission (Tg CH_4/year)
Coal mining	15–45
Coal combustion	1–30
Extraction of oil	5–30
Extraction and use of natural gas	25–50
Total fossil	46–155
Sewage treatment plants	15–80
Sanitary landfills	20–70
Domestic animals	65–100
Animal waste	20–30
Rice paddies	20–100
Biomass burning	20–80
Total biospheric	160–460
Total	206–615

Nitrous oxide

An inventory of global emissions of nitrous oxide (N_2O) is presented in Table 3.6. The primary sources of N_2O emissions are industrial processes and nitrogen fertilizers, with the latter estimated to be responsible for 50–75% of total emissions. There is a factor of 2 uncertainty in the total global emission, with a larger relative uncertainty for some terms in the inventory. Catalytic converters are increasingly used in automobiles in order to reduce NO_x emissions, but they increase N_2O emissions. Emissions from fertilizer use (on cultivated soils) vary widely depending on the nature of the fertilizer and especially depending on the soil water content and the timing of fertilization.

Table 3.6 Estimates of global anthropogenic N_2O emissions, as summarized by Prather *et al.* (1995)

Nitrous oxide source	Emission (Tg N/year)
Cultivated soils	1.8–5.3
Biomass burning	0.2–1.0
Industrial sources	0.7–1.8
Cattle and feed lots	0.2–0.5
Total	2.9–8.6

Halocarbons

The term "halocarbon" refers to compounds containing either chlorine, bromine, or fluorine, together with carbon, many of which act as powerful GHGs. These include the chlorofluorocarbons (CFCs), the hydrochloro-fluorocarbons (HCFCs), and the hydrofluorocarbons (HFCs). The chlorine- and bromine-containing halocarbons are involved in the depletion of the stratospheric ozone layer. Halocarbons are produced by the chemical indus-try and are used in a variety of applications, mostly involving refrigeration and cooling. The emission of halocarbons to the atmosphere depends on, first, the amount that is used in a given application (such as air condition-ing), and second, the extent to which leakage to the atmosphere can be prevented. In the case of air conditioning, the amount used depends on the design and efficiency of the cooling equipment and on the overall cooling requirements of the building, while the leakage depends on how well main-tained the equipment is and how servicing and disposal of cooling equipment are carried out.

Aerosols and aerosol precursors

Aerosols can be classified as "primary" and "secondary". Primary aerosols are those that are directly emitted, while secondary aerosols are produced through the chemical transformation of precursor gases. Table 3.7 summarizes esti-mates of the total anthropogenic emissions of primary aerosols or aerosol precursors, or of the rate of production of secondary aerosols, during the 1980s. The main aerosols of interest are: (i) sulphate (SO_4^{2-}) aerosols, (ii) soot or "black carbon" aerosols, which are partially absorbing, and (iii) non-absorb-ing organic carbon aerosols. The uncertainty in sulphur emissions, at ±25% of the central estimate, is comparatively small; for some other aerosol types, the uncertainty is a factor of 5.

As noted in Section 2.5, sulphate aerosols are produced from the chemical transformation of SO_2 that is emitted during the combustion of S-containing coal, the refining of S-containing oil, and the smelting of minerals such as zinc, copper, and lead, which are mixed with sulphur. Table 3.8 gives the range of sulphur emission factors for various processes based on present practice. Sulphur emissions are currently being reduced in many developed countries, owing to their impact on regional air pollution and their role in acid rain formation.

Carbon (or carbonaceous) aerosols occur in two forms: as soot, which is directly emitted during incomplete combustion of fossil fuels or organic matter, and as reflective organic compounds that are produced photochemi-cally from volatile (off-gased) organic compounds. The latter can be subdivided into organic gases produced during the burning of biomass, and as non-methane hydrocarbons (NMHCs) that are released to the atmosphere largely through the use of refined petroleum products (such as gasoline).

Table 3.7 Estimates of the global anthropogenic emissions of major aerosol types during the 1980s, as summarized by Jonas *et al.* (1995), except where indicated

Aerosol type	Estimated flux (Tg/year) Range	Estimated flux (Tg/year) Preferred
Primary		
Industrial dust	40–130	100
Soot, total	5–25	10
from fossil fuels[a]	7–8	
from biomass burning[a]	5–6	
Secondary		
SO_x emission (Tg S/year)	66–94[b]	67[c]
SO_4 production from SO_x (Tg S/year)[d]	40–60	47
NO_x emission (Tg N/year)	21–30[e]	25[f]
Nitrates from NO_x (Tg N/year)[g]	6–15	12
Organic matter from biomass burning	50–140	80
Organic matter from non-methane hydrocarbons (NMHCs)	5–25	10

[a] Based on Cooke and Wilson (1996) and Liousse *et al.* (1996)

[b] This is the range for estimates of fossil fuel SO_x emissions for the inventory years 1985 and 1986, as given by Benkovitz *et al.* (1996), plus 2 Tg S/yr to account for biomass-related emissions. On average, about 80% of emitted SO_x is converted to sulphate before being removed from the atmosphere.

[c] This is the preferred estimate of Benkovitz *et al.* (1996), including 2 Tg S/yr from biomass burning.

[d] The estimates by Jonas *et al.* (1995) are in mass of SO_4, but have been converted here to mass of S to permit comparison with the SO_x emission, which is also given in mass of S. In the S chemistry model of Roelofs *et al.* (1998), about 80% of emitted SO_2 is converted to sulphate, the rest being directly deposited at the Earth's surface.

[e] This is the range for estimates of total anthropogenic NO_x emissions for the inventory years 1985 and 1986, as given by Benkovitz *et al.* (1996) and IEA (1991b), the latter being shown in disaggregated form in Table 3.11.

[f] This is the preferred estimate of 21 Tg N/yr of Benkovitz *et al.* (1996), which largely excludes biomass emissions, plus a middle estimate (from Table 3.11) for emissions due to biomass burning of 4 Tg N/yr.

[g] The estimates by Jonas *et al.* (1995) are in mass of NO_2, but have been converted here to mass of N to permit comparison with the NO_x emission, which is also given in mass of N.

Precursors of tropospheric ozone

Tropospheric ozone is produced as a byproduct of the oxidation of CO and CH_4 if the concentration of NO is sufficiently high (Section 2.8). It is also produced as a byproduct of the oxidation of NMHCs in the presence of adequate NO. Emissions of CO, CH_4, NMHCs, and NO_x (= NO + NO_2) therefore all lead to the production of O_3 in the troposphere, so these gases are ozone precursors. Tables 3.9–3.11 present estimates of the global emissions of CO, NMHCs, and NO_x (CH_4 emission estimates were given in Table 3.5). There are uncertainties in the energy-related emissions of CO and NO_x of probably at least ±50%, while the uncertainty associated with biomass emissions is even larger. Biomass burning is a significant source of NO_x and especially CO emissions, while transportation is the largest fossil fuel-related source of CO and NO_x emissions on a global basis. Because the atmospheric lifespan of both ozone and its precursors is rather short (days to weeks), there are marked variations in the concentration of tropospheric ozone.

Table 3.8 Sulphur emission factors as (kg S/kg C) for the use of coal and oil, or as (kg S/tonne metal) for the smelting of metals. Source: see references below

Process	Uncontrolled emissions	Present-day controlled emissions	Best controlled emissions
Coal combustion	0.2–2.2 kg S/GJ[a,b]	0.1–0.5[c,d]	0.0
Oil refining	0.1–0.4 kg S/GJ[e,f,g]	≤0.06[f]	0.0–0.03[f]
Primary copper smelting	1060 kg/ton[h]	⎫ 90–95% reduction[i]	⎫ 99% reduction[i]
Secondary copper smelting	225 kg/ton[h]		
Primary lead smelting	149 kg/ton[h]		
Secondary lead smelting	42.6 kg/ton[h]		
Primary zinc smelting	490 kg/ton[h]	⎭	

[a] EIA (1998).

[b] Range of sulphur emission factors is for coal with a sulphur content ranging from 0.5% to 5%.

[c] EPA (1998).

[d] Princiotta (1991).

[e] Naturalgas.com: *The Environment: Acid Rain*, at http://naturalgas.com/environment/acid/html.

[f] Syncrude Canada (1998).

[g] This range of emission factors is for fuel oil with sulphur content of 0.5% to 2%.

[h] Spiro *et al.* (1992).

[i] Charles Ferguson, Vice President, Environment, Health & Safety, Inco Limited.

Table 3.9 Estimated total global anthropogenic emissions of CO in 1992, including both direct emissions and emissions arising from the anthropogenic production of CH_4 and NMHCs (non-methane hydrocarbons). Taken from Lelieveld *et al.* (1998)

Source	Emission (Tg C/yr)
Direct emissions	
Energy use	230
Biomass burning	170
Indirect emissions	
Oxidation of CH4	243[a]
Oxidation of NMHCs	50

[a] Derived from an assumed anthropogenic CH_4 emission of 304 Tg C/yr and a CO production rate of 0.8 moles per mole of emitted CH_4.

Summary of emissions of trace gases from changes in land use

Reference has already been made to emissions of trace gases in association with the burning of biomass. Biomass burning is, however, only one of three or four steps in the transformation of tropical forests to other land uses. The first step is the clearing (cutting) of standing vegetation. This is followed by the burning of felled vegetation, planting of grain and root crops, and finally, conversion to pasture with periodic burning to control regrowth. Emissions of trace gases, or

Table 3.10 Estimated inventory of global emissions (Tg C/year) of various non-methane hydrocarbons, as given by Wang *et al.* (1998a)

Species	Anthropogenic				Natural	
	Industrial	Biomass	Oxidation	Total	vegetation	Total
Ethane	6.3	2.5	0	8.8	0	8.8
Propane	6.8	1.0	0	7.8	0	7.8
$\geq C_4$ alkanes	30	0	0	30	0	30
$\geq C_3$ alkenes	10.4	12.6	0	23	0	23
Isoprene	0	0	0	0	597	597
Acetone	1.0	8.9	12.4	22.3	15	37.3
Total	54.5	25.0	12.4	91.9	612	703.9

Table 3.11 Estimated inventory of global anthropogenic emissions of NO_x in 1986 (teragrams N per year), as given in Tables A-9 and A-10 of IEA (1991b) for all sources except biomass burning, which is based on Crutzen and Andreae (1990) and Andreae (1995). N emissions given by IEA (1991b) are reported in terms of the mass of NO_x, but have been converted here to mass of N by assuming that all of the NO_x is NO_2

Source	OECD[a]	CPE[b]	LDC[c]	World
Industrial	1.2	1.4	0.5	3.2
Residential/commercial	0.6	0.9	0.6	2.1
Transportation	6.0	1.9	2.4	10.3
Electricity generation	3.2	3.4	1.1	7.7
Other transformation	0.4	0.5	0.3	1.2
Total energy-related	11.5	8.1	4.9	24.5
Biomass burning	—	—	—	2–8

a OECD = Organization for Economic Cooperation and Development.
b CPE = Centrally Planned Economies.
c LDC = Less Developed Countries.

changes in the natural emissions or sinks of trace gases, are associated with each of these steps. Table 3.12 presents a qualitative summary of the effects of clearing, burning, and agricultural or pastoral use of former tropical forests on emissions of CO, CH_4, NMHCs, NO_x, N_2O, and aerosol particles or aerosol precursors. Biomass burning causes large emissions of these substances, most of which serve as ozone precursors. A large fraction (about 15%) of the carbon that is released to the atmosphere is released as CO. This CO is eventually oxidized to CO_2 (on a time scale of about 20 days). Up to 2% of the carbon is released as CH_4. This CH_4 is also eventually oxidized to CO_2 (on a time scale of about 10 years), but in the interim, each CH_4 molecule is 26 times as effective as a CO_2 molecule in contributing to the greenhouse effect (Table 2.2). Substantial quantities of sooty (absorbing) and non-absorbing carbon aerosols are also released, particularly from smouldering woody material. Agricultural

Table 3.12 Effects of deforestation on trace gas and aerosol emissions, based on Keller *et al.* (1991)

	CO	CH$_4$	NMHC	NO$_x$	N$_2$O	Aerosols
Clearing	?	–	–	+	+	?
Burning	++	++	++	++	+	++
Use	?	+	–	++	++	?

Key: – decreased emissions

     + increased emissions

     ++ greatly increased emissions

use of land (after it has been cleared and burned) has its largest effects on N$_2$O and NO$_x$ emissions, which tend to increase sharply. The increase in NO$_x$ emissions is related to changes in soil temperature and moisture and in foliage density. The main cause of increased N$_2$O emissions is the application of N fertilizers although, as noted above, the magnitude of the emissions depends strongly on soil water content and the timing of fertilization.

Comparison of natural and anthropogenic emissions

Table 3.13 compares natural and anthropogenic emissions for those gases where such a comparison is meaningful; that is, for all the gases discussed so far except CO$_2$. Anthropogenic emissions are a significant fraction of natural emissions for CH$_4$, SO$_2$, N$_2$O, CO, NO$_x$, and many aerosols. All GHGs or aerosol precursors except CO$_2$ are removed from the atmosphere entirely, or almost entirely, through chemical destruction (either by reacting with OH or through photodissociation). The rate of removal is proportional to the concentration within the atmosphere (although not necessarily linearly proportional, as discussed in Section 7.9). A 10% increase in the total emission would eventually lead to a roughly 10% increase in concentration. In the case of CO$_2$, the main removal processes are photosynthesis and inflow into the ocean, not irreversible chemical destruction. As the removal processes increase in strength (owing to an initial increase in concentration), the return flows or natural emissions also increase. Thus, as photosynthesis increases, the supply of organic matter increases, and this allows the respiration flux to increase. Similarly, as CO$_2$ flows into the ocean, the back-pressure of CO$_2$ in the water increases (quite rapidly, as discussed in Section 7.5), so the outflow also increases. The removal process becomes permanent only as CO$_2$ works its way into the deep ocean, a process that takes on the order of 1000 years. This makes the atmospheric CO$_2$ concentration particularly sensitive to additional emissions. To illustrate this point, note that total natural emissions are about 160 Gt C yr^{-1} (about 60 Gt C yr^{-1} due to respiration of plant detritus and soil carbon, and about 100 Gt C y^{r-1} due to gaseous diffusion from the oceans). Total anthropogenic emissions are about 5% of this (6-8 Gt C yr^{-1}) but, if sustained at this level, would eventually lead to CO$_2$ concentrations *many times* the initial concentration.

Table 3.13 Comparison of total natural and anthropogenic emissions for gases that are removed from the atmosphere through irreversible chemical destruction. Also given are nitrogen fixation rates on land and the estimated natural and anthropogenic emissions of aerosols or aerosol precursors. As explained in the text, it is not meaningful to directly compare natural and anthropogenic emissions of CO_2. Except where indicated, anthropogenic emissions are taken from other tables in this chapter, while natural emissions are taken from Chapter 2

Gas	Units for emission	Natural emissions	Anthropogenic emissions
CH_4	Tg CH_4 per year	190 ± 70[a]	340 ± 160[a]
N_2O	Tg N per year	6–12[b]	3.9–7.6
SO_x	Tg S per year	8[c]	60–100[d]
All S species	Tg S per year	16–31[c]	60–100
CO	Tg C per year	360[d]	700
NO_x	Tg N per year	0.5–10[e]	21–30
N fixation	Tg N per year	150	40
C aerosols:			
Soot	Tg C per year	0	5–25
Organic C from fires	Tg C per year	?	50–140
Organic C from NMHCs	Tg C per year	55	5–25

[a] From Lelieveld *et al.* (1998). In combining natural and anthropogenic emissions, the total source must equal 600 ± 80 Tg/yr. The uncertainty range for anthropogenic emissions given here largely overlaps the range given in Table 3.5 from the summation of individual anthropogenic sources.

[b] From Prather *et al.* (1995).

[c] Pham *et al.* (1996); Rasch *et al.* (1999).

[d] From Lelieveld *et al.* (1998).

[e] From Table 12.3 of Jaffe (1992).

3.2 Variation in GHG concentrations and in GHG and aerosol emissions up to the present

Concentrations of CO_2, CH_4, and N_2O

Systematic atmospheric measurements of the concentration of CO_2, CH_4, and N_2O began in 1958, 1978, and 1976, respectively. Concentrations prior to these years have been determined by measurements of the concentrations of these gases in air bubbles from ice cores extracted from Antarctica and Greenland. Figure 3.1 shows the variation in the atmospheric concentration of CO_2, CH_4, and N_2O since 1800 based on a composite of ice core data from Antarctica and directly measured concentrations at the South Pole.

Figure 3.2 provides a more detailed look at the change in CO_2 concentration from 1960 to 1998, as measured at the Mauna Loa observatory in Hawaii. A seasonal variation in the CO_2 concentration is seen, superimposed on an overall upward trend. The seasonal cycle is due to the drawdown of atmospheric CO_2 in NH spring and summer, when photosynthesis exceeds

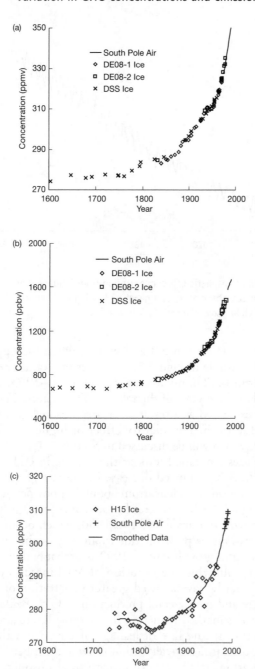

Figure 3.1 Variation in the observed atmospheric concentration of (a) CO_2, (b) CH_4, and (c) N_2O from 1600 to 1995. The data sources are as follows: ice core CO_2, Etheridge *et al.* (1996); ice core CH_4, Etheridge *et al.* (1992); ice core and atmospheric N_2O, Machida *et al.* (1995); atmospheric CO_2 and N_2O, *Trends '97* (web site: http://cdiac.esd.ornl.gov). Dave Etheridge and Toshinobu Machida kindly provided their data in electronic form.

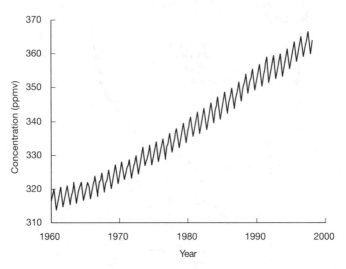

Figure 3.2 Variation in atmospheric CO_2 concentration from January 1960 to December 1997, as measured at Mauna Loa observatory and given by Keeling and Whorf (1998) and available at the web site: http://cdiac.esd.ornl.gov.

respiration, and a net release in fall and winter, when the opposite occurs. Although barely discernible in Figure 3.2, the seasonal variation has increased by about 15% (from 6.52 ppmv averaged over 1960–1963, to 7.65 ppmv averaged over the last three years of the record). The increase in seasonal variation can be partly explained as a result of greater seasonal growth (and hence decay) of plants due to the stimulatory effect of higher atmospheric CO_2 (Houghton, 1987), which will be discussed in Section 7.1.

Figure 3.3 provides a detailed look at the change in CH_4 concentration from 1984 to 1996. A pronounced decrease in the rate of growth of atmospheric CH_4 concentration is evident, from about 15 ppbv per year to less than 5 ppbv per year (Figure 3.3(b)). Several factors contributed to these changes. The decrease in growth rate from 1984 to 1990, which occurred in spite of increasing CH_4 emissions, is probably due to an increase in OH (see Section 4.6). The abrupt drop in growth rate by 1992 is probably due to the decrease in stratospheric O_3 following the eruption of Mt Pinatubo in June 1991, which would temporarily have allowed greater penetration of UV radiation into the troposphere and even greater production of OH (Bekki et al., 1994). The continued but less marked reduction in the growth rate after 1992 could be due to a reduction in the rate of leakage of CH_4 from the natural gas pipeline system in the former Soviet Union (Law and Nisbet, 1996). Finally, year-to-year fluctuations in the rate of growth of atmospheric CH_4 can be partly explained by the dependence of both the natural emission of CH_4 (from wetlands) and its removal (by reaction with OH) on temperature (Bekki and Law, 1997). It should also be noted that, with constant emissions, the atmospheric CH_4 concentration would level off after a time comparable to the atmospheric lifespan of CH_4, which is on the order of 10 years.

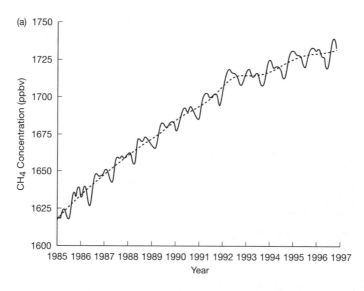

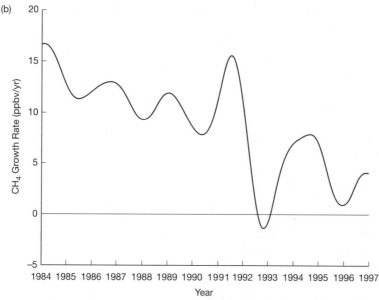

Figure 3.3 (a) Globally averaged variation in atmospheric CH_4 concentration from 1984 to 1997. (b) Rate of growth in average atmospheric CH_4 concentration from 1984 to 1997. Reproduced from Dlugokencky *et al.* (1998).

CO_2 emissions

Plate 1(a) shows the variation in total global emissions of CO_2 due to the burning of fossil fuels and broken down by fuel type, while Plate 1(b) shows an estimate of net emissions due to all land use changes and broken down by ecosystem. The land use emission estimates are from Houghton (1998), and reach 2.2 Gt C yr^{-1} by 1990, which is near the high end of the 1.6 ± 1.0 Gt C yr^{-1} range given earlier. According to Plate 1(b), land use emissions increased slowly from 1870 to 1940, then rose steadily and sharply after 1940. Emissions from non-tropical forest ecosystems (i.e., from developed countries) exceeded tropical emissions until about 1940, at which point tropical emissions began to increase sharply while emissions from other ecosystems decreased. By 1990, essentially all of the land use emissions were from the tropics.

The rate of increase of atmospheric CO_2 during the past 120 years has increased more or less in proportion to the increase in the annual anthropogenic emissions. This match is one of several lines of evidence that allow us to state with complete certainty that the increase in atmospheric CO_2 since the Industrial Revolution was caused by human emissions. These lines of evidence are summarized in Box 3.1.

Box 3.1 How can we be certain that human CO_2 emissions have caused the recent increase in atmospheric CO_2?

One of the few things that is known with absolute certainty is that the increase in the atmospheric CO_2 concentration during the past two centuries is a direct result of human emissions of CO_2, and not the result of some natural fluctuations that occurred coincidentally during and after the Industrial Revolution. The evidence supporting this statement is as follows.

- The recent increase in atmospheric CO_2 far exceeds the bounds of natural variability experienced during the preceding 250,000 years.
- The rate of increase during the past century has increased in step with increasing anthropogenic emissions.
- The atmospheric CO_2 concentration is slightly higher in the northern hemisphere, where the bulk of human emissions occur, and the interhemispheric concentration difference has been increasing over time.
- The atmospheric ratios of ^{14}C to ^{12}C and of ^{13}C to ^{12}C have declined during the past two centuries in the manner expected if human emissions are responsible for the buildup of atmospheric CO_2.
- Various lines of evidence indicate that the undisturbed terrestrial biosphere and the oceans – the only possible alternative candidates to explain the CO_2 buildup – have both been sinks rather than sources of atmospheric CO_2.

In spite of this overwhelming evidence, there are still sometimes suggestions in the fringe or pseudo-scientific literature (e.g., Seitz, 1994) that the current CO_2 increase might not be due to human emissions. Such suggestions are based on a gross misunderstanding of how the oceanic part of the carbon cycle works – a topic that is discussed in great detail in Chapter 7.

Methane emissions

Plate 2 shows the historical variation in anthropogenic emissions of methane, as estimated by Stern and Kaufmann (1996a). The 1990 emission of about 370 Tg CH_4 falls in the middle of the uncertainty range of 206–615 Tg CH_4 given in Table 3.5. According to Plate 2, fossil fuels contributed 20% of the total anthropogenic emission, which is also in accord with Table 3.5. There is considerable uncertainty in the CH_4 emission estimates. Nevertheless, some qualitative features of Plate 2 are most probably correct: that there were already substantial anthropogenic emissions by 1870; that most of the growth since then has been due to emissions from fossil fuels, livestock, and landfills; that emissions from biomass burning and rice paddies have also grown since 1870, but by a smaller factor; and that the rate of emission growth was larger after 1940 than before 1940.

Sulphur emissions

Past concentrations of aerosols cannot be measured. However, since the life-span of aerosols in the troposphere is so short (a few days), the concentration responds essentially instantaneously to changes in emissions. Although the global mean concentration does not vary exactly linearly with the global emission, variations in the estimated emission can be used as a rough proxy for the variation in concentration. Figure 3.4 shows the past variation in anthropogenic sulphur emissions as estimated by Stern and Kaufmann (1996b), based on inventories of fossil fuel use and metal smelting. The 1985 emission was 83 Tg S according to Figure 3.4; alternative estimates of the total emission for 1985 range from 65.0 Tg S (Chin et al., 1996) to 94.4 Tg S (Pham et al., 1996). According to Figure 3.4, global sulphur emissions rose rapidly until 1910, experienced less rapid growth from 1910 to 1940, rose sharply between 1940 and the late 1980s, and then decreased. From 1970 to the mid-1990s, OECD[2] emissions fell by about 40%, while non-OECD emissions rose sharply. The decrease in global emissions after 1990 is largely due to a decrease in non-OECD emissions (in particular, falling emissions in eastern Europe and the former Soviet Union).

A critical factor in determining anthropogenic effects on future climate is the ratio of sulphur to carbon emissions due to the burning of fossil fuels alone. This ratio is shown in Figure 3.5, based on the data shown in Plate 1(a) and Figure 3.4. As can be seen from Figure 3.5, there has been a steady downward trend in the S:C emission ratio. There is no reason not to expect this trend to continue for a very long time.

[2] OECD = Organization for Economic Cooperation and Development. The OECD countries roughly correspond to the industrialized countries, exclusive of the former Soviet Union.

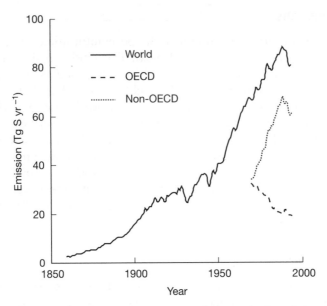

Figure 3.4 Historical variation in global anthropogenic S emission (Tg S yr^{-1}), as estimated by Stern and Kaufmann (1996b) and given at the web site http://cres.anu.edu.au/~dstern/anzsee/datasite.html. Also shown, since 1970, are separate emissions from OECD and non-OECD countries. This breakdown is not available for times prior to 1970.

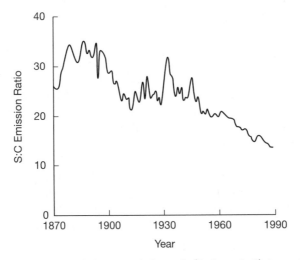

Figure 3.5 Variation in the S:C mass emission ratio (Tg S per Gt C) due to the combustion of fossil fuels, based on the CO_2 emission data given in Plate 1(a) and the S emission estimates of Figure 3.4.

Questions for further thought

1. Why are efforts to control acid rain also relevant to future climatic change?
2. In what ways will efforts to reduce traditional air pollution problems affect future climate?
3. In what ways will efforts to preserve the world's remaining forests affect future climate?
4. How does the link between emissions and atmospheric concentration differ for sulphur and methane?

Chapter 4

Observed changes in the climate system and sea level during the recent past

In this chapter the various lines of evidence indicating that the Earth's climate is beginning to warm are reviewed. In addition, evidence concerning changes in a number of other climate system variables is discussed. Observations pertaining to changes in the amount of water vapour in the atmosphere are presented, because of the importance of water vapour as a GHG and its positive feedback effect on climatic change. Evidence concerning changes in the amount of ozone in the stratosphere and troposphere is discussed, because considerable effort and uncertainty are involved in deducing ozone trends owing to the great spatial variation in ozone concentration (this is why ozone trends were not discussed along with those of well-mixed GHGs in the preceding chapter). Evidence concerning changes in the OH and CO concentration of the atmosphere is discussed, as changes in these two gases serve as indicators of the changing chemical state of the atmosphere. Finally, evidence concerning the current rate of sea level rise and efforts to explain it are discussed because of their relevance to projected future sea level rise.

4.1 Indicators of global mean temperature change

There are a number of indicators of changes in the global average temperature: records of sea surface temperature (SST), marine air temperature, and surface-air temperature over land; measurements of temperatures in the free atmosphere made from balloons; measurements of the variation of temperature with depth in boreholes; satellite measurements of microwave emission; and records of tree ring width and density.

Surface and surface–air temperature records

Thermometer-based measurements of air temperature were systematically recorded at a number of sites in Europe and North America as far back as 1760, but the set of observing sites did not attain sufficient geographic coverage to permit a rough computation of the global average land temperature until the mid-19th century. Sea surface temperatures (SSTs) and marine air temperatures (MATs) have been systematically collected by merchant ships since the mid-19th century, but these temperature records are restricted to the routes followed by commercial ships, so that there are relatively few data from large regions in the southern hemisphere even today. Land-based, marine air, and sea-surface temperature datasets all require various corrections to account for changing conditions and measurement techniques.

Land-based temperature records need to be screened for inconsistencies that can arise from changes in the measurement site, instruments, instrument shelters, or the way that monthly averages were computed, or due to the growth of cities around the sites (which would lead to a spurious warming trend due to a growing urban heat island). Effects of urbanization can be estimated and removed either by comparison of individual urban stations with nearby rural stations (as in Jones *et al.*, 1990), or by comparing results when cities with populations over 100,000 are included or excluded from the analysis (as done by Hansen and Lebedeff, 1987).

There are a number of reasons why SST and MAT data need to be corrected. Heating of the ship deck by the sun will alter daytime MATs. This problem can be eliminated by basing marine air temperature trends on nighttime marine air temperatures (NMATs) only. Up to 1941, most SST measurements were made in buckets of sea water that were hoisted onto the ship deck, while after 1941, most measurements were made at the engine water intakes. Figure 4.1 shows photographs of some of the different kinds of buckets that were in use in the 19th century. Between about 1856 and 1910, there was a transition from wooden buckets to canvas buckets, but the timing of the transition is uncertain. Evaporative cooling of the bucket after having been placed on the ship deck would have occurred in both cases, but to a much greater extent with canvas buckets. There was a gradual transition from sailing ships to steamships between 1876 and 1910, which altered the height of the ship deck and the speed of the ship. A change in the height of ship decks would have affected MATs, while the change in ship speed would have affected the evaporative cooling of the buckets.

The simplest way to correct the marine data is to adjust the datasets to achieve consistency with other. This is the approach that was followed by Jones *et al.* (1986). The first step is to adjust the MAT (or NMAT) anomalies so that the marine and adjacent land temperature anomalies move in parallel at time scales of a decade and longer. This requires that the land data have already been corrected to ensure internal consistency. In the second step, the SST anomalies are adjusted so that they move in parallel with the adjusted MAT (or NMAT) anomalies. Jones *et al.* (1991) refer to this as the *a posteriori* approach. An alternative approach is to compute the required correction based on first principles and one's understanding of the causes of errors – what Jones *et al.* (1991) call the *a priori* approach. This has been used to correct both NMAT and SST anomalies.

The important information in the global temperature analysis is the year-by-year deviation of the temperature from the average during some reference period. These temperature deviations are referred to as "anomalies", although this is quite an inappropriate term. Plates 3 and 4 show uncorrected global mean NMAT and SST anomalies, respectively, and the anomalies as corrected

Figure 4.1 Various kinds of buckets used in the 19th and early 20th centuries for measuring sea surface temperature. Reproduced from Folland and Parker (1995).

by Folland *et al.* (1984), Bottomley *et al.* (1990), and Parker *et al.* (1995). Successive correction attempts have resulted in markedly different reconstructed SST anomalies for the period prior to 1900. The most recently calculated correction increases steadily from about 0.1°C in 1856 to 0.4°C in 1940, as the proportion of canvas buckets steadily increased, then abruptly drops to zero in 1941 when engine intake measurements began.

Figure 4.2 presents a composite of the corrected global mean SST and land-surface air temperature anomalies for the period 1861–1998. Temperature deviations for individual years, as well as a running average (using the five years centred on the current year), are shown. An overall warming of about 0.6–0.7°C is seen, with the warming occurring in two steps: between 1910–1940 and after 1970. 1998 was by far the warmest year on record, with a temperature about 0.6°C warmer than the average for the period 1961–1990.

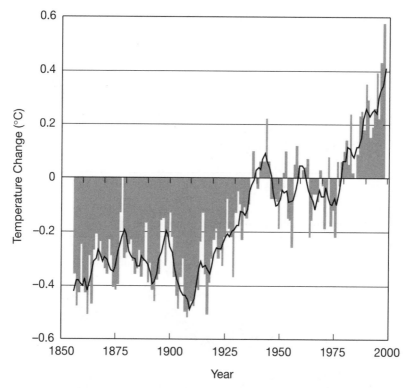

Figure 4.2 Combined annual land-surface air and SST anomalies (°C) for 1860 to 1998, as provided by the UK Meteorological Office Web Site (http://www.meto.gov.uk).

The number of observation sites and their geographical range varied substantially over the time period represented by Figure 4.2. The question arises as to what effect these changes might have had on the inferred hemispheric and global mean temperature trends. This question has been addressed in a variety of ways, and a consistent result is that changes in the observation network had, at most, a small effect on the overall trend during the past century. However, the ranking of individual years with similar mean temperatures does depend on the spatial distribution of the observation grid. Short-term oscillations involve changes in winds that produce limited spatial correlations, while longer-term oscillations are more related – it appears – to external factors that give better spatial correlations. Thus, the longer the time scale under consideration, the greater the correlation between spatially separated monitoring sites and the fewer the number of evenly spaced points that are required in order to get a good global average.

Figure 4.3 shows the uncertainty in the global, NH, and SH combined land air–SST temperature anomalies due to changes in the observational network, as computed by Jones *et al.* (1997). Using the central estimate ± 1 standard deviation, the global mean warming from 1851 to 1995 could be as small as 0.5°C or as large as 0.8°C (i.e., 0.65 ± 0.15°C). However, if one applies the

(a) Global mean decadal temperature anomaly

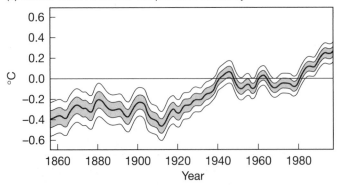

(b) Northern hemisphere decadal temperature anomaly

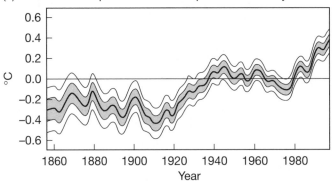

(c) Southern hemisphere decadal temperature anomaly

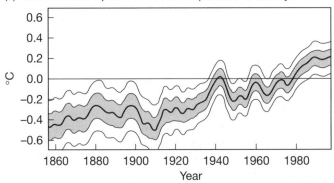

Figure 4.3 Combined land-air and SST temperature anomalies (°C) for (a) the globe, (b) northern hemisphere, and (c) southern hemisphere, with the central estimate (heavy lines), ±1 standard deviation (shaded), and ±2 standard deviations (thin lines). Reproduced from Jones *et al.* (1997).

estimated errors by random amounts each year, estimates the resultant temperature trend, and then repeats this process 1000 times to determine the distribution of temperature trends so computed, one obtains an estimated trend and uncertainty of about $0.65 \pm 0.05°C$ (Phil Jones, personal communication, November 1998).[1] The exact uncertainty depends on how correlated one assumes the errors to be from one year to the next.

Balloon-based temperature records

A global network of balloon-launching sites was established in May 1958, during the International Geophysical Year. The balloons contain sensors that directly measure temperature, relative humidity, and pressure as they rise through the atmosphere to the altitude where the balloons burst (typically around 20 km). The original balloons were tracked optically and were referred to as "radiosondes"; the term "rawindsonde" is used if the device can give wind information in some way (as is now the case). The initial network consisted of about 540 stations that reported once daily at 0000 Greenwich Mean Time. Beginning in 1963, twice-daily reporting, at 0000 and 1200 GMT, began to be important. By the 1970s the network grew to 700–800 stations with twice-daily reporting.

There are number of factors that can cause spurious trends in the balloon-based temperature data, as discussed by Gaffen (1994) and Parker and Cox (1995). These include changes in the temperature sensor used, changes in the way that the effects of solar heating of the instrument package are accounted for, and changes in the heights and temperatures at which the balloons burst. The adjustments to the measured temperature to account for solar radiation are well below 1°C near the surface but increase with increasing height to about 20°C in the upper stratosphere; the occasional changes in the computed correction in the stratosphere are on the order of a few tenths of a degree Celsius (Gaffen, 1994). Such changes are significant compared to the long-term stratospheric cooling of 1.5°C inferred from the raw rawindsonde data. Parker and Cox (1995) state that the early balloons were more likely to burst in cold conditions, creating a spurious cooling trend in the lower stratosphere of unknown magnitude as the balloons were strengthened.

Table 4.1 gives the trends in NH and SH mean temperature for the period 1964–1989 at various pressure heights in the atmosphere, as deduced from balloon data, along with the trends in surface temperature. The global mean trends from the balloon 950 mb data and from surface measurements differ by almost a factor of two. The balloon data indicate a significantly greater near-surface warming in the SH than in the NH. Nearly identical warming is seen for the 850–300 mb layer. A weak cooling is seen in the upper troposphere (300–100 mb) and a strong cooling in the stratosphere. The stratospheric results should be regarded with suspicion but, as discussed next, can be compared with independently derived trend estimates.

---

[1] This procedure is referred to as a Monte Carlo estimation technique.

Table 4.1 Trends in hemispheric mean air temperature for the period 1964–1989, based on measurements taken from balloons and as analysed by Oort and Liu (1993). Also given are the trends computed from the combined surface air–sea surface temperature data set that is available from the Climate Research Institute, University of East Anglia (http://www.cru.uea.ac.uk)

Atmospheric layer	Temperature trend (°C/10 yr)		
	NH	SH	Global
CRU–surface	0.12 ± 0.03	0.16 ± 0.02	0.13 ± 0.02
Surface–950 mb	0.13 ± 0.06	0.32 ± 0.07	0.24 ± 0.06
850–300 mb	0.20 ± 0.10	0.23 ± 0.09	0.22 ± 0.09
300–100 mb	−0.03 ± 0.10	−0.11 ± 0.11	−0.07 ± 0.09
100–50 mb	−0.38 ± 0.14	−0.43 ± 0.16	−0.40 ± 0.12

Satellite-based proxy records

Microwave sensors, known as microwave sounding units (MSUs), have been placed aboard satellites and are able to measure microwave radiation (at a wavelength of 4 mm) emitted by O_2. Since the atmospheric concentration of O_2 is constant in space and in time, variations in the emitted radiation should be directly related to variations in the temperature of emission. Two different channels sample the radiation at slightly different frequencies, which correspond to emissions in different regions of the atmosphere: MSU channel 2R samples the middle and lower troposphere (mainly the 850–300 mb layer), while channel 4 largely samples the stratosphere (mainly the 50–100 mb layer).

The MSU data have the distinct advantage of providing global coverage, unlike the rawindsonde and surface-based measurements, both of which are concentrated over land and in the northern hemisphere. However, numerous problems and errors with the MSU data have been repeatedly discovered, resulting in frequent revisions to the computed temperature trends. This is in contrast to the temperature trends from the land-based record over the period for which MSU data are available (1979 to the present), which are quite stable. The most recently discovered error arises from the fact that the height of each satellite has gradually decreased over time due to the frictional effect of the outer atmosphere. This introduced a spurious cooling trend of 0.12°C/decade (Wentz and Schabel, 1998). In addition, the MSU record is still rather short, which makes the accurate determination of long-term trends difficult.

Figure 4.4 shows the variation in global mean temperature deduced for MSU-2R and MSU-4, along with the variation – since 1964 – in the rawindsonde temperatures from comparable layers in the atmosphere, and in surface air temperature. The MSU-2R temperatures shown in Figure 4.4(b) are those of Wentz and Schabel (1998), in which the effects of changing satellite altitude have been corrected for. Two rawindsonde time series are shown – one by Oort and Liu (1993), based on the global rawindsonde network of 700–800 stations, and the other by Angell (1997), based on a carefully chosen subset of

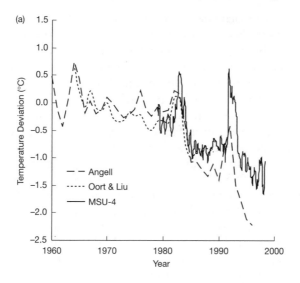

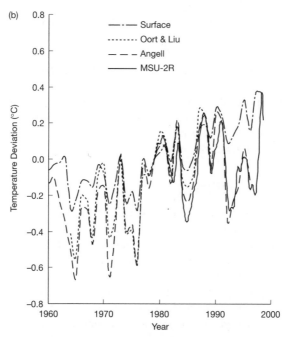

Figure 4.4 (a) Comparison of temperature records for the stratosphere (50–100 mb). Shown are the MSU-4 record (solid line) and the balloon-based records of Oort and Liu (1993) and Angell (1997). (b) Comparison of tropospheric and surface temperature records. Shown are the MSU-2R record for the troposphere (mainly the 850–300 mb layer) from Wentz and Schabel (1998), the balloon-based records of Oort and Liu (1993) and Angell (1997) for the 800–350 mb layer, and the surface record prepared at the University of East Anglia (http://www.cru.uea.ac.uk). The Angell (1997) data were obtained from the Carbon Dioxide Information and Analysis Center (http://cdiac.esd.ornl.gov/ftp/ndp008).

63 stations. In order to facilitate intercomparison, all of the temperature series shown in Figure 4.4(b) have been shifted to give zero deviation in 1979. There is good agreement between the balloon-based and corrected satellite-based trends. The global mean tropospheric temperatures experienced a much greater decrease in response to the eruption of Mt Pinatubo (in mid-1991) than did the surface temperature, but both rebounded to the same amount by the end of 1998. This is quite reasonable, given the moderating effect of the large thermal mass of the ocean surface layer compared to the atmosphere.

The trend in tropospheric temperature, after correcting for the aforementioned effects of decreasing satellite altitude, is a warming of 0.07°C/decade over the period 1979–1995 (Wentz and Schabel, 1998). By comparison, Jones (1994) computed temperature trends for the period 1979–1993 of 0.10°C/decade based on rawindsonde data for the 850–300 mb layer, and 0.17°C/decade based on combined land-surface air and SST data, respectively.[2] Part of the difference between MSU/rawindsonde trends and surface-based temperature trends could be related to a real difference in the surface and mid-tropospheric temperature changes, or it could be partly related to uneven sampling of surface air temperature.

Turning to the stratosphere, the trend for MSU-4 from 1979 through to April 1997 is a cooling of 0.49°C per decade (Christy *et al.*, 1998). The rawindsonde data of Angell (1997) show a much stronger cooling over this period (Figure 4.4(a)). The Oort and Liu (1993) dataset agrees with the MSU-4 data during the period of overlap (1979–1989). This agreement does not validate the rawindsonde trend prior to 1979, because most of the known procedural or instrument changes that would have caused spurious trends occurred prior to 1979. The real trend prior to 1979 may have been smaller than shown in Figure 4.4(a).

Ground temperature profiles

Variations in surface temperature over time will propagate into the Earth's sub-surface with some attenuation, producing a vertical variation in temperature at any given time that reflects the temporal variation in surface temperature. It is therefore possible to reconstruct surface temperature variations from borehole records of the temperature variation with depth. For such reconstructions to have any climatic relevance, it must first be determined that the vegetation at the site has not been altered over the time interval of interest, as changes in vegetation can cause changes in the surface temperature. Analysis of borehole records from widespread sites that pass this test indicates that surface warming has occurred during the 20th century at almost all sites examined (Nicholls *et al.*, 1996, Section 3.2.5.2; Harris and Chapman, 1997).

---

[2] These trends were computed after removing the warming effect of an El Niño at the beginning of the record, and the cooling effect of Mt Pinatubo just after 1991. Prior to adjustment, the surface trend was 0.1°C/decade. The effect of Mt Pinatubo would be smaller for the time period (1979–1995) considered by Wentz and Schabel (1998), so removal of the effect of Mt Pinatubo would have had a smaller effect on the trend.

Changes in alpine glaciers, seasonal snow cover, and sea ice

In almost all parts of the world where mountain glaciers occur, retreat of the glaciers has occurred during the 20th century. Although the mass budget of glaciers depends on the amount of snowfall, as well as on temperature, the observed retreat is consistent with a warming in alpine regions of 0.6–1.0°C (Oerlemans, 1994).

Figure 4.5 shows the variation in the annual and seasonal extent of snow-cover in the NH (excluding Greenland) and the variation in surface air temperature in regions where snowcover has decreased. The annual mean extent of snowcover on land in the NH decreased by about 10% during the period 1972–1992, with the largest decreases in spring and autumn (Groisman *et al.*, 1994). The temperature and snowcover variations are closely related. Figure 4.6 shows the inferred variation in the extent of sea ice in the northern hemisphere as computed by Maslanik *et al.* (1996). A clear down-ward trend in the area of sea ice is evident. Consistent with the trend of decreasing extent of sea ice, Smith (1998) finds that the annual number of days in which melting of sea ice occurs increased at a rate of 5.3 days (8%) per decade over the period 1979–1996.

Figure 4.7 shows the variation in the extent of SH sea ice between 1978 and 1996, as computed by Stammerjohn and Smith (1997) based on satellite data. There is no trend in winter and spring, and a weak upward trend in summer and autumn. However, evidence from whaling ships indicates that the total area of SH sea ice decreased in extent by about 25% between the mid-1950s and the early 1970s (de la Mare, 1997) – prior to the period of satellite observations.

In addition to sea ice, the Antarctic continent is fringed by ice shelves that are fed by glacier flow from the land-based ice masses. Vaughan and Doake (1996) report that the five northernmost ice shelves on the Antarctic Peninsula retreated dramatically during the past 50 years, while those further south show no clear trend. Comparison with air temperature records – which show a warming of about 2.5°C since 1945 – indicates that there is a temperature threshold near a mean annual temperature of –5°C, with shelf retreat begin-ning once this threshold is crossed.

Changes in the length of the NH growing season

Further evidence of a recent warming of climate over NH land is provided by satellite observations of plant photosynthetic activity, which indicate that the growing season north of 45°N has increased by 12 ± 4 days during the period 1981–1991 (Myneni *et al.*, 1997).

Changes in the distribution of plants and animals

Epstein *et al.* (1998) summarize evidence indicating that the distributions of a number of plant and animal species have changed in a manner that is consis-tent with a warming of the climate. For example, upward displacements of

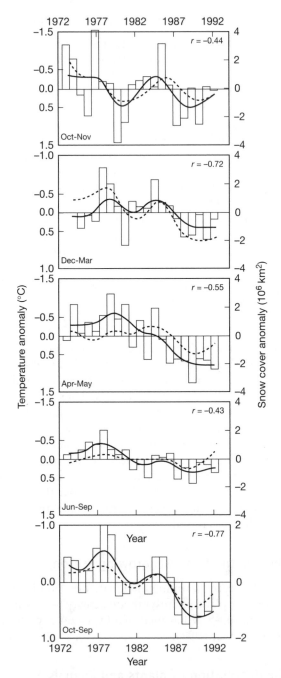

Figure 4.5 Variation in the mean annual and seasonal extent of snowcover on land in the NH (bars, yearly values; solid lines, smoothed values), and of surface temperature (dashed lines) over regions of transient snow cover. Reproduced from Nicholls *et al.* (1996).

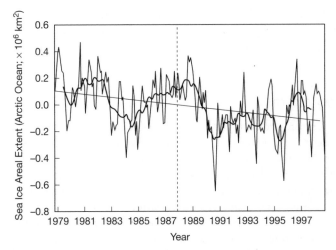

Figure 4.6 Variation in the extent of sea ice during the period 1979 to September 1998 as measured by satellites for the northern hemisphere. This is an update of Maslanik *et al.* (1996), kindly provided by James Maslanik.

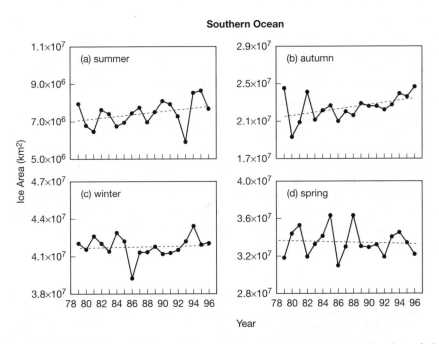

Figure 4.7 Variation in seasonal sea ice area in the southern hemisphere during the period 1979–1994 as inferred from satellite data. This is an update of Stammerjohn and Smith (1997), using a datafile kindly provided by Sharon Stammerjohn.

plant distributions have been documented on 30 alpine peaks, the distribution of a number of butterfly species has shifted poleward, and mosquito-borne diseases (such as malaria and dengue fever) are being reported at higher elevations than before.

Long proxy records

Proxy temperature indicators at a number of locations allow us to extend temperature records back several centuries. Mann *et al.* (1998) estimated the variation in NH mean annual temperature since AD 1400 based on such indicators as tree-ring width and density, the chemical composition and annual growth rate in corals, the characteristics of annual layers in ice cores, and long historical records. The reconstructed variation in NH mean annual temperature is shown in Plate 5 along with error bars that show the range within which the true annual temperature variations are estimated to fall with a probability of 96%. Also shown is the instrumental record of NH average temperature change since 1900, which agrees very closely with the reconstructed temperature change during the period of overlap. The reconstruction indicates that the 20th century has been warmer than at any time since 1400.

4.2 Geographic, seasonal, and diurnal patterns of warming _____

The magnitude of surface warming during the recent past varied with location, season, and time of day. Plate 6 shows the pattern of mean annual temperature warming from 1950–1959 to 1990–1998. Warming is most pronounced at high latitudes, but there are also broad regions (particularly over the oceans) where cooling has occurred. In most regions where warming has occurred, the largest warming is in winter and spring.

There has been a marked asymmetry in the warming between night and day. Analysis of non-urban land stations over the period 1950–1993 and representing about 50% of the global land area indicates that nighttime minimum temperatures have warmed more than twice as fast as daytime maximum temperatures – at a rate of 1.79°C per 100 years compared to 0.84°C per 100 years. Consequently, the diurnal temperature range (DTR) has decreased at a rate of 0.79°C per 100 years (Easterling *et al.*, 1997).[3] Regions and seasons showing the strongest reduction of diurnal temperature range are associated with statistically significant increases in cloudiness.

---

[3] The trend in DTR does not equal the trend in maximum temperature minus the trend in the minimum temperature for two reasons. (1) Discontinuities (due to changes in instrumentation) in all three temperature series at each station were corrected only if the discontinuity exceeded a specified threshold. Sometimes discontinuities in DTR exceeded threshold while those for the maximum and minimum temperature did not. (2) Some stations have only maximum or minimum temperatures, so different sets of stations go into the computation of the mean trends for the three temperature series.

4.3 Trends in precipitation, storminess, and El Niño _____

A climatic parameter that is every bit as important as temperature is the amount (and distribution) of precipitation, but precipitation – whether as rain or snow – is much harder to measure than temperature. Difficulties arise not only from the tendency of precipitation gauges to underestimate the amount of rainfall due to the effects of turbulence around the gauge created by the gauge itself, but also because precipitation varies strongly from one point to the next. It is therefore difficult to derive meaningful areal-mean precipitation amounts unless a very dense network of precipitation gauges is used – a condition that is rarely satisfied. However, as long as changes in the precipitation gauges are properly accounted for, it should be possible to correctly deduce at least the direction in which precipitation is changing, even if the absolute areal-mean precipitation amounts still contain significant errors.

Mean annual precipitation

There are two global precipitation datasets: "Hulme" (Hulme, 1991; Hulme *et al.*, 1994), and the "Global Historical Climate Network" (GHCN), described by Vose *et al.* (1992) and Eischeid *et al.* (1995). There is no clear trend in global precipitation, but regional trends are evident, as illustrated in Figure 4.8 for the periods 1955–1974 to 1975–1994 and 1900–1994. Precipitation appears to have increased in most regions except southern Europe, the Sahel region of Africa, Indochina, Japan, and Chile.

Heavy precipitation

Karl *et al.* (1995) found a trend towards a greater proportion of rainfall occurring as heavy-rainfall events in the USA (1911–1992), the former Soviet Union (1935–1989), and China (1952–1989). Suppiah and Hennessy (1998) report trends of both increasing and decreasing total rainfall and heavy rainfall in Australia, depending on the region.

Snowfall in Antarctica and Greenland

Snowfall over Antarctica appears to have increased by 5–20% in recent decades, based on the changing thickness of annual layers in Antarctic ice cores and estimates of the convergence of atmospheric water vapour fluxes over Antarctica (Walsh, 1995). In contrast, Fischer *et al.* (1998) report evidence from an ice core in central Greenland that the rate of snow accumulation has decreased by 20% during the past 50 years. This is consistent with evidence – discussed in Section 10.1 – of an inverse relationship between regional temperature and snowfall in Greenland during the Holocene Epoch (the last 10,000 years).

(a)

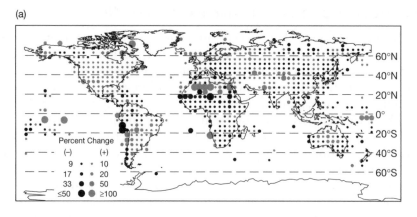

(b)

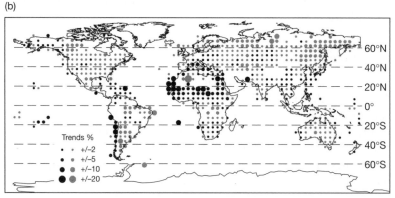

Figure 4.8 (a) Changes in precipitation over land from 1955–1974 to 1975–1994, expressed as a percentage of the 1955–1974 mean. (b) Changes in precipitation over land from 1900 to 1994, expressed as a percentage of the 1961–1990 or 1951–1980 mean (depending on the region). Reproduced from Nicholls *et al.* (1996).

Tropical cyclones

There appears to be a weak downward trend in the intensity and frequency of the most intense hurricanes in the Atlantic Ocean (Landsea *et al.*, 1996). Figure 4.9 shows the average peak intensity reached by all hurricanes in each year during 1944–1995, and the maximum intensity reached by the most intense hurricane of the year. Both peak and average intensities show a downward trend. Figure 4.10 shows the variation in the frequency of intense Atlantic hurricanes (having sustained surface winds of at least $50 \, \mathrm{m \, s^{-1}}$ at some point) and in the remaining hurricanes; a downward trend is evident for intense hurricanes, with the 1995 hurricane season standing out as an unusually active season. In the case of the western North Pacific Ocean, the frequency of hurricanes decreased from the early 1960s to about 1979, and has been increasing since then but has not yet reached previous activity levels (Chan and Shi, 1996).

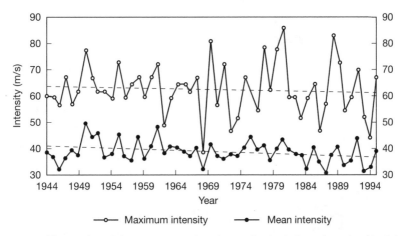

Figure 4.9 Time series of the mean annual peak sustained windspeed attained in Atlantic hurricanes and of the windspeed attained by the strongest hurricane. Reproduced from Landsea *et al.* (1996).

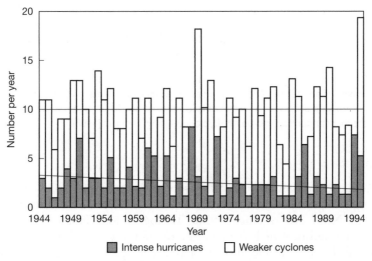

Figure 4.10 Time series of annual frequency of intense hurricanes in the Atlantic basin, and of all remaining hurricanes. Intense hurricanes are defined as those with a maximum sustained windspeed of at least 50 m s⁻¹. Reproduced from Landsea *et al.* (1996).

Extratropical cyclones

Figure 4.11 shows the 11-year running mean of the number of intense winter extratropical cyclones in the NH for the period 1899–1991, as deduced by Lambert (1996) based on the analysis of sea level pressure maps. The Pacific data prior to 1930 are not considered to be reliable. There was little or no trend in the frequency of winter cyclones between 1930–1970, followed by a sharp increase after 1970 in the Pacific Ocean sector.

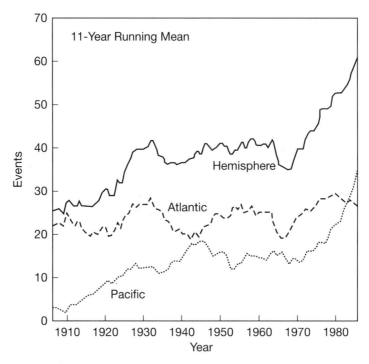

Figure 4.11 Eleven-year running means of the number of intense winter extratropical cyclones for the whole NH and for the Atlantic and Pacific sectors. An intense cyclone is defined as the occurrence of a grid point with a mean sea level pressure of 970 mb or less which is lower than each of the four surrounding grid points. Reproduced from Lambert (1996).

El Niño

El Niño-like conditions existed in the tropical Pacific from 1990 to June 1995, the longest such El Niño on record. There has also been a trend for more frequent El Niño events since 1976. Trenberth and Hoar (1996, 1997) performed a statistical analysis of an El Niño index for the period 1882–1997, and found that the likelihood of the recent changes occurring due to natural variability alone, given the previous record, is only one in 2000 years. The long El Niño of 1990–1995 was followed by one of the strongest El Niños on record during 1997–1998, which reinforces the viewpoint that the statistical characteristics of El Niño have changed during the last 20 years.

4.4 Changes in the amount of water vapour in the atmosphere

Determining the way in which the amount and vertical distribution of water vapour in the atmosphere have varied as the surface climate varied over the last few decades would be of great value in assessing the long-term sensitivity of climate to human and natural perturbations. This is because water vapour is a greenhouse gas that increases in amount as the climate warms, thereby amplifying the initial warming.

The total amount of water vapour in an overhead column is referred to as the *precipitable water*. It is given as the depth of liquid water that would form if all of the water vapour were to be condensed out. Current datasets can be used to determine how the amount of precipitable water has changed with surface temperature over the last few decades. However, current datasets are inadequate for assessing long-term trends in the amount of water vapour in the upper troposphere (above the 500 mb pressure level). Water vapour in this region does not make a large contribution to precipitable water, but has a disproportionately large effect on the natural greenhouse effect.

There are three separate issues pertaining to observations of the vertical distribution of water vapour in the atmosphere: determining the *climatology* or long-term average of the water vapour distribution; determining the vertical structure of *interannual variations* in the amount of water vapour in the atmosphere; and determining the vertical structure of *long-term trends* in the amount of water vapour in the atmosphere. Long-term trends would be most useful for assessing climatic feedbacks, but are the least reliable of the three owing to changes in instrumentation. However, the climatology of the water vapour distribution provides insight into the relative importance of different processes that operate in the real atmosphere and that would therefore be involved in climatic feedbacks; the observed climatology also serves as one test of the water vapour processes in computer simulation models that are used to forecast long-term changes in climate. The computed climatology of the water vapour distribution is less sensitive to changes in instrumentation than are computed trends. The nature of the interannual variability in atmospheric water vapour as given by observed datasets is also more reliable than the long-term trends, since it is less affected by instrumentation problems, but might not be fully applicable to long-term climatic change. Thus, the trends that are most relevant to long-term climatic change are the least reliable, while the most reliable information is of questionable utility for projecting future climatic change.

Table 4.2 summarizes a number of recent analyses of the decadal and longer trends in atmospheric water vapour. Efforts have been made to account for known changes in instruments and data processing procedures, so the trends in the total precipitable water and in the amount of water vapour in the lower troposphere should be reliable. The results summarized in Table 4.2 consistently show that both total precipitable water and the water vapour amount in the lower troposphere have increased during the past 1–2 decades, at the same time that the global mean temperature has increased. Most of the instances where the amount of precipitable water has decreased coincide with areas where temperatures have also decreased.

Also shown in Table 4.2 (as items 6–7) are results from balloon-based, satellite-based, and ground-based instruments designed specially for measuring the amount of water vapour in the stratosphere. Although these records are shorter than the others, they also show an upward trend in water vapour amount, albeit with considerable scatter. The mixing ratio trend of 0.13–0.15 ppmv/yr

Table 4.2 Trends in atmospheric water vapour, based on analysis of rawindsonde data

Source	Years	Layer	Quantity	Region	Trend
1	1965–1984	500–700 mb	W	EIP	Increasing at 10 stations (mostly by 3–5%/decade), slight decrease at the others
2	1973–1990	1000–300 mb	W	TWP	6%/decade
		1000–850 mb	RH	TWP	4%/decade
3	1973–1990	1000–200 mb	W	Global	Generally increasing, typically by 5–10%/decade at low latitudes
4	1973–1993	1000–500 mb	W	USA	3–4%/decade
5	1981–1993	10–26 km	W	B	3–10%/decade in 2 km layers
6	1991–1997	40–60 km	MR	Global	0.13 ppmv/year
7	1992–1997	40–60 km	MR	C, NZ	0.15 ppmv/year

Sources:
    1 = Hense *et al.* (1988)
    2 = Gutzler (1992)
    3 = Gaffen *et al.* (1992)
    4 = Elliott and Angell (1997)
    5 = Oltmans and Hofmann (1995)
  6,7 = Nedoluha *et al.* (1998)

Quantities:
   W = precipitable water
  RH = relative humidity
  MR = mixing ratio

Regions:
  EIP = equatorial Indian and Pacific Oceans
  TWP = tropical western Pacific Ocean
    B = Boulder (Colorado), USA
    C = Table Mountain (California), USA
   NZ = Lauder, New Zealand

is on an initial water vapour amount of about 5.5 ppmv. About half of the upward trend can be explained by the observed increase in stratospheric CH_4 (which is oxidized to H_2O), while the other half can be explained by an increase in temperature at the tropical tropopause by 0.1 K/yr (this would permit more water vapour to enter the stratosphere by increasing the saturation vapour pressure, which governs how much moisture that rising tropospheric air can hold).

4.5 Changes in the amount of ozone in the atmosphere

The total amount of ozone in the atmospheric column at a given location can be measured using ground-based instrumentation, using instruments launched on balloons, and using satellite-based remote sensors. In some cases, information on the vertical distribution (or profile) of ozone can also be obtained. Surface concentrations can also be directly measured from ground-based instruments.

There is a worldwide network of about 16 sites at which balloons are launched, about once per week; these balloons reach heights in excess of 30 km before bursting. However, the results are usable only to a height of 27 km, owing to instrument limitations (WMO, 1998). Observations at most sites began between 1966 and 1970.

There are four major satellite-based ozone datasets. Satellite observations of ozone began in October 1978 with the Total Ozone Mapping Spectrometer (TOMS) on board the Nimbus 7 satellite, and have continued with TOMS instruments on other satellites. The raw data have been re-analysed several times as computational procedures and supplementary data have improved. TOMS gives vertically integrated ozone amounts. The second satellite-based ozone dataset is from the Stratospheric Aerosol and Gas Experiment (SAGE) instruments. SAGE gives the vertical profile of O_3, with 1 km resolution, from cloud tops to a height of 55 km (SAGE I) or 65 km (SAGE II). Interference with aerosols makes it difficult to determine trends accurately below a height of 20 km, and significant problems still remain (WMO, 1998). The third satellite-based ozone dataset is from the Solar Backscatter Ultraviolet (SBUV) instrument. SBUV was launched on Nimbus 7 (along with TOMS) and collected data from November 1978 to June 1990. It can determine the vertical profile of ozone in the stratosphere and above. A second SBUV instrument (SBUV-2) was launched on board the NOAA-11 satellite, and collected data from January 1989 to October 1994. The fourth satellite-based ozone dataset is from the TIROS-N operational vertical sounders (TOVS) on NOAA[4] polar orbiting satellites. The deduced trends are most representative of the low stratosphere (WMO, 1998).

Trends in total ozone

The easiest ozone trends to decipher are the trends in the total amount of ozone in the atmospheric column at a given location. Reported trends in the amount of ozone are based on statistical regression models that attempt to remove the effects of variations in solar activity (as indicated by sunspot number) and in stratospheric winds, which cause short-term variations in the amount and distribution of ozone. Figure 4.12 shows the variation with latitude in the trend in total mean annual overhead ozone, as measured by ground-based Dobson spectrophotometers and as measured by the TOMS satellites. The trends are given as percent per decade, but were computed over the period January 1979 to December 1997. Trends near the equator are not statistically significant, while the losses reach 6%/decade at high latitudes in the SH and 4%/decade at high latitudes in the NH. The typical uncertainty is ±25%. There has been a noticeable slowing in the rate of O_3 loss in the 1990s that parallels the slowdown in the rate of increase of the chlorine loading in the stratosphere (WMO, 1998).

---

[4] National Oceanographic and Atmospheric Administration, US government.

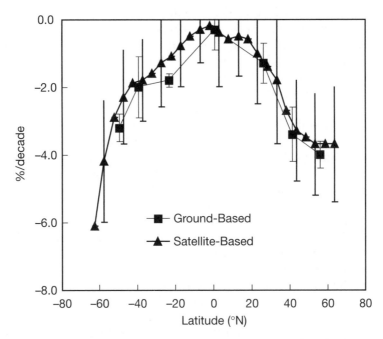

Figure 4.12 Variation with latitude in the trend in annual mean total overhead ozone as measured by Dobson spectrophotometers and by TOMS satellite instruments. Also given are estimated uncertainties, but to avoid clutter, uncertainties are given for only every other latitude for the satellite-based data. Based on data in Tables 4.5 and 4.6 of WMO (1998).

Trends in stratospheric ozone

Trends in the amount of ozone in individual atmospheric layers, such as the upper and lower stratosphere and the troposphere, are much less reliable than the trends in total ozone. This is unfortunate, since the climatic impact of changes in the amount of ozone depends critically on the vertical variation in the changes. Of particular importance are changes in the amount of ozone near the tropopause, in both the lower stratosphere and upper troposphere. Figure 4.13 shows the vertical distribution of the trend in ozone at latitudes 30°N–50°N as inferred from SAGE, SBUV, and Umkehr data. Particularly strong disagreements are evident in the critical 15–25 km layer, next to the tropopause.

Trends in tropospheric ozone: changes since the Industrial Revolution

Surface ozone observations were made at various sites in Europe as far back as 1839 (when ozone was first discovered). The earliest observations made at high-altitude sites under pollution-free conditions are from the Pic du Midi

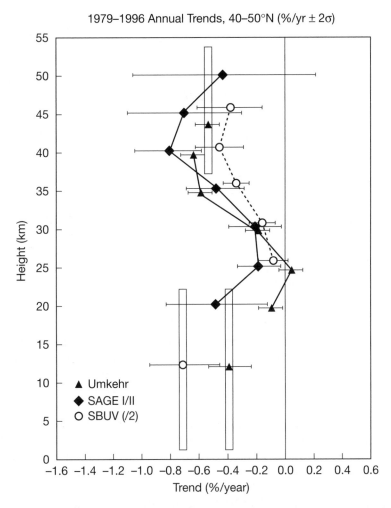

Figure 4.13 Vertical variation of the trend in the annual mean amount of ozone as measured by SBUV, SAGE, and Umkehr instruments. Reproduced from WMO (1998).

observatory for the period 1874–1909. The Pic du Midi observatory is near the Atlantic coast of France at an elevation of 3000 m, so the measurements made there should be representative of a broad region of the free troposphere in the NH mid-latitudes. Marenco *et al.* (1994) review the history of ozone measurements and the efforts made to permit comparison between values obtained using early and modern methods. Figure 4.14 shows the original Pic du Midi observations along with more recent observations from the Pic du Midi and other high-altitude sites in Europe. There is a very clear increase in the ozone concentration in the free troposphere over Europe, by a factor of 5, during the past 120 years. The present-day concentration is around 50 ppbv.

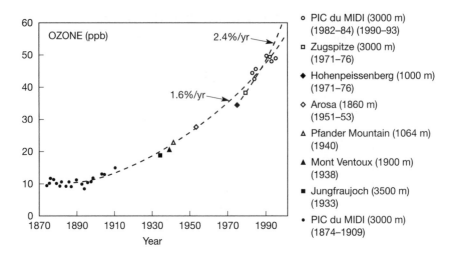

Figure 4.14 Historical measurements of the concentration of ozone at high-altitude sites in Europe, as analysed and compiled by Marenco *et al.* (1994).

Surface ozone was also measured at Montevideo (Uruguay) during 1883–1885 and at Cordoba (Argentina) during 1886–1892. Sandroni *et al.* (1992) analysed these data and concluded that the surface ozone concentration in clean air in the southern hemisphere has roughly doubled (from annual mean values of about 10 ppbv to about 20 ppbv) during the past century. A second indication of the increase in the amount of ozone in the SH relies on observed seasonal variations. There are large seasonal variations in the amount of tropospheric ozone in the tropics, related to seasonal variations in biomass burning which, combined with regional differences in the amount of ozone, can be used to set lower and upper limits to the increase in tropospheric ozone in the tropics due to biomass burning. Using this approach, Portmann *et al.* (1997) inferred that the vertically integrated increase in tropospheric ozone during the biomass burning season (August–September–October) is from two to almost four times that over the tropical South Atlantic region.

Trends in tropospheric ozone: changes in recent decades

Recent trends in the amount of ozone in the troposphere can be determined (i) from ozonesonde data, (ii) by subtracting the amount of ozone above the tropopause, as determined from SAGE data, from the concurrent amount of total ozone as determined from TOMS data, or (iii) by comparing TOMS total ozone amounts over the high Andes (altitude about 6 km) and the adjacent Pacific Ocean. Using the latter method, Jiang and Yung (1996) were able to deduce the amount of ozone in the lower 6 km of the atmosphere, which accounts for most of the tropospheric ozone.

The ozonesonde data and satellite data together indicate that the amount of ozone in the troposphere has been increasing during the last 1–2 decades in most regions. The changes in the amount of ozone in the troposphere during the past two decades, as measured by ozonesondes, result in changes in the total column amount of ozone ranging from a decrease of 1.1%/decade to an increase of up to 1.5%/decade, with most stations producing an increase of 0.0–0.8%/decade (Logan, 1994; her Table 10). Next to South America, Jiang and Yung (1996) derived an upward trend during the period 1979–1992 of 14.8 ± 4.0% per decade (a cumulative increase of 14–24%). The increase in tropospheric ozone and decrease in stratospheric ozone have opposing effects on surface climate.

4.6 Very recent trends in the atmospheric OH and CO concentrations.

Both CO and OH have very short atmospheric lifespans, that for CO ranging from 20 to 50 days, and that for OH ranging from 1 to 10 seconds. For this reason, their concentration varies considerably spatially (and also temporally in the case of OH). This makes the determination of trends very difficult. As discussed in Section 2.8 and summarized in Table 2.1, OH is crucial to the chemistry of many climatically important gases (and to the production of sulphate aerosols). Since CO is a major sink for OH, it is also indirectly of great climatic importance. Here, evidence concerning recent trends in the OH and CO content of the atmosphere is briefly discussed.

Hydroxyl

The global distribution of OH cannot be measured in practice, so trends cannot be measured either. Rather, recent trends in the concentration of OH are inferred from 3-D atmospheric chemistry models that are constrained to match the observed spatio-temporal variation in the concentration of methylchloroform (MCF, CH_3CCl_3). MCF is chosen because it is solely of anthropogenic origin, its rate of emission is well known, and reaction with OH is the dominant removal process. In this way, Krol *et al.* (1998) deduced an OH trend of a 0.5 ± 0.6% per year increase for the period 1978–1993. A very tentative breakdown of the possible factors responsible for this increase is given in Table 4.3. The two largest factors contributing to the increase appear to be the decrease in stratospheric O_3 (which permits more ultraviolet radiation to reach the troposphere) and the increase in NO_x emissions. The recent decrease in CO emissions (see below) and the increase in the amount of water vapour in the tropical atmosphere also appear to be important. It should be stressed, however, that the inferred increase in OH is still rather tentative. Prinn *et al.* (1995) deduced an essentially zero trend for the period 1978–1994.

Table 4.3 Tentative breakdown of the causes of the inferred global mean increase in the tropospheric OH concentration by $7.5 \pm 9.0\%$ from 1978 to 1993, based on simulations with a 3-D atmospheric chemistry model by Krol *et al.* (1998). The difference between the last two lines is a result of non-linearities in the atmospheric chemistry

Causal factor	% Change in OH
11% increase in the CH_4 concentration	−1.1
6.5% decrease in the CO concentration	+1.7
Loss of stratospheric O_3	+2.0
0.2°C temperature increase	+0.1
10% increase in tropical H_2O	+1.7
10% increase in NO_x emissions	+2.0
Sum of the above	+6.4
Simultaneous imposition of the above	+6.0

Carbon monoxide

A 3-D model of atmospheric chemistry indicates that the CO concentration has increased by a factor of 2 throughout much of the NH troposphere, and by a factor of 1.5 throughout much of the SH troposphere, from pre-industrial times to the present (Crutzen and Zimmermann, 1991). As for recent trends, a transition from increasing to decreasing global mean CO concentration occurred in the late 1980s, such that the average concentration decreased by 2.0% per year in the NH and 3.0% per year in the SH after 1990 (Novelli *et al.*, 1998). This decrease appears to be due to a combination of reduced CO emissions from automobiles (due to cleaner vehicles) and from biomass burning (due to a recent reduction in the amount burnt).

4.7 Sea level

Changes in sea level measured at a coastal measuring site could be due to some combination of vertical movements in the edge of the continent and of changes in the volume of the oceans. The measured change in sea level is thus the local change relative to the continental margin, not the absolute change. Vertical movements can occur due to the normal geological compaction of deltaic sediments, the withdrawal of groundwater from coastal aquifers, the uplift associated with colliding tectonic plates, or the ongoing postglacial rebound associated with deglaciation at the end of the last ice age. Changes in ocean volume can occur due to human-induced withdrawals or addition of water to the ocean (a direct effect), the melting of small glaciers and ice caps, and the expansion of ocean water as it warms. Only the latter two are indicative of global-scale climatic change, so the effects of local vertical movements of the land surface and of non-climate-related changes in ocean volume need to be subtracted from the observed records, and the geographically disparate records combined to produce a meaningful global average. There are over 1400 tide-gauge stations worldwide, but only about 150 stations have records longer than 50 years, mostly located in the northern hemisphere. As in surface temperature trends (Section 4.1), the longer the time scale under consideration, the greater the correlation between spatially separated sampling sites, and the fewer the number of sites that are needed to get a globally representative trend.

Recent rates of sea level rise

In order to reliably estimate trends in sea level, the effect of the rebound of the Earth's crust after the retreat of glaciers from the last ice age has to be taken into account. This so-called isostatic rebound and the compensating sinking elsewhere extend far beyond the areas that were covered by ice. A geophysical model, calibrated using ^{14}C-dated paleo-shorelines, is needed to determine local glacial–isostatic effects and to subtract these effects from the tide-gauge records. Using this approach, Peltier and Tushingham (1989) deduced a rate of sea level rise in the 20th century of 2.4 ± 0.9 mm/yr. Douglas (1991) used the regional isostatic rebound rates obtained by Peltier and Tushingham (1989) but a different tide-gauge dataset, in which some of the stations used by Peltier and Tushingham were rejected because of local tectonic effects. The resulting dataset consisted of 21 stations distributed throughout the NH with records of 60 years or longer. He obtained a sea level rise during 1880–1980 of 1.8 ± 0.1 mm/year. The very small uncertainty range in this estimate (compared to that of Peltier and Tushingham) is due to the very narrow distribution in the sea level trends from individual stations in the dataset that Douglas (1991) used. Later, Douglas (1997) was able to add three more stations with records of 60 years or longer, all from the SH, and obtained the same rate of sea level rise (1.8 ± 0.1 mm/yr). By comparison, the rate of sea level rise for several centuries prior to the last century was an order of magnitude smaller (Douglas, 1995).

Factors contributing to the recent sea level rise

An appreciation of the difficulty in accurately projecting future changes in sea level – which is the subject of Chapter 10 – can be gained by examining attempts to explain the recent 2 mm/yr rate of sea level rise above the long-term rate. Table 4.4 gives estimates of the various contributions to the cumulative sea level rise during the past century or so. Direct anthropogenic effects, which are discussed in some detail by Sahagian et al. (1994) and Gornitz et al. (1997), could be responsible for a fall in sea level of around 1 cm, with an uncertainty of perhaps ±40%. Melting of small glaciers and the Greenland ice cap may have added 4–7 cm to sea level. Thermal expansion of ocean water may have added another 2–5 cm. These effects give a sea level rise

Table 4.4 Estimates of the contribution of various processes to the change in sea level during the past century. Estimates of direct effects were obtained by integrating the time-dependent annual rates of change in sea level given in Figure 2 of Gornitz et al. (1997), except where indicated. Alternative estimates of some direct anthropogenic effects by Sahagian et al. (1994) are given in brackets

Process	Contribution (cm)
Direct anthropogenic effects	
Permanent removal of water from aquifers	0.98 (1.23)[a]
Deforestation and loss of soil moisture	0.67 (0.34)
Reduction in the extent of wetlands	<0.01 (0.13)
Storage behind dams	−0.38 (−0.52)
Deep infiltration of water behind dams	−1.71
Deep infiltration of irrigation water	−1.82
Increase in atmospheric water vapour content	−0.54[b]
Effects of urbanization	1.51
Combustion of fossil fuels	0.29[c]
Total direct anthropogenic effects	−1.00
Induced by climatic change	
Melting of small glaciers	2.7 ± 0.9[d]
Melting of Greenland	3.0 ± 0.7[d]
Changes in Antarctic ice mass	0 ± 14[e]
Thermal expansion of sea water	2.2–5.1[f]
Total without Antarctica	6.3–12.4
Total with Antarctica	−8 to 26
Observed	18 ± 1

[a] The estimate by Sahagian et al. (1994) includes loss of water from the Caspian and Aral seas, which were not considered by Gornitz et al. (1997). Eliminating these contributions brings their total to 0.88 cm, which agrees better with Gornitz et al. (1997).

[b] This number is mean to represent the effect of enhanced evaporation from reservoirs and irrigated fields in increasing the storage of water vapour in the atmosphere, but seems to be excessive. Given that the global mean precipitable water content of the atmosphere is only 2.5 cm, this implies 25% increase. Based on the observed trends in atmospheric water vapour reported in Table 4.2, this seems more appropriate for the increase induced by climatic warming at the global scale during the past century.

[c] Gornitz et al. (1997) estimate a current rate of sea level rise of 0.021 mm/yr due to the current fossil fuel use at a carbon release rate of 6 Gt per year. Given that the cumulative fossil fuel carbon release to 1990 was about 140 Gt, this corresponds to a cumulative rise of sea level of 0.29 cm (neglecting the effect of changes in the fuel mix on the CO_2:H_2) emission ratio).

[d] Zuo and Oerlemans (1997).

[e] Warrick et al. (1996).

[f] de Wolde et al. (1995).

of 5–12 cm, which is significantly less than the estimated observed sea level rise of 17–19 cm. The shortfall can be made up by changes in the mass balance of Antarctica, which is independently estimated to have caused a change in sea level of 0 ± 14 cm.

In order to move the central estimate of the directly computed increase in sea level up to the central estimate of the observed increase, we must assume that Antarctica has experienced a net loss of mass as the climate has warmed. This is contrary to observations (discussed in Section 4.4) indicating that an increase in Antarctic snowfall has occurred in recent decades, and contrary to expectations of what should happen in the 21st century. However, recent studies (summarized by Oppenheimer, 1998) indicate high rates of melting at the base of the floating ice shelves that fringe West Antarctica. Combined with estimates of a positive mass balance elsewhere, this could produce a sea level rise of 1.9 mm per year if the loss of floating ice is compensated by outflow from the land-based ice.

Acceleration of the sea level rise

The warming of the past 150 years – which in many long records stands out as unusual during the past 1000 years or more – should have led to an acceleration in the rate of sea level rise for reasons explained above. Douglas (1995) reviews several lines of evidence, from sites in Europe and on the US east coast, that indicate that the rate of sea level rise prior to the mid-19th century was substantially smaller than during the past 150 years (2 mm/yr), or even indistinguishable from zero. Thus, an acceleration *has* occurred at some time in the recent past. However, analysis of a globally representative set of tidal gauge records shows that no statistically significant acceleration in sea level rise has occurred (Douglas, 1992). Indeed, the lack of any observed acceleration is a much more robust result than any given estimate of sea level rise, since uncertain constant contributions to the apparent sea level rise (such as postglacial rebound) fall out of the calculation (Douglas, personal communication, 1998).

4.8 Summary

The trends identified in this chapter are summarized in Table 4.5. Direct measures of surface or near-surface atmospheric temperature indicate that the climate has warmed during the 20th century. This conclusion is reinforced by trends in many variables that are directly influenced by temperature: alpine glaciers have retreated worldwide, the extent of snowcover and sea ice has decreased (at least in the NH), and the growing season has increased in length. Precipitation and the proportion of total precipitation as heavy precipitation have increased in many regions of the world. Changes in the frequency of tropical and extratropical cyclones, and in the intensity of tropical cyclones, have also occurred, but since projections of how these should change as the climate warms are contradictory (as will be discussed in Sections 8.5 and 8.6), these changes cannot be taken as being consistent (or inconsistent) with a

Table 4.5 Summary of trends in observed climatic variables that are discussed in this chapter

Variable	Analysis period	Trend or change
Surface air temperature and SST	1851–1995	$0.65 \pm 0.05°C$
Diurnal temperature range	1950–1993	$-0.79°C$/century
Stratospheric temperature	1979–1995	$-0.9°C$
Extent of snowcover in NH	1972–1992	10% decrease in annual mean
Extent of sea ice in NH	1973–1994	Downward since 1977
Extent of sea ice in SH	1973–1994	No change, possible decrease between mid-1950s and early 1970s
Ice shelves off Antarctic Peninsula	Last 50 years	Dramatic retreat in north
Alpine glaciers	20th century	Widespread retreat
Length of NH growing season	1981–1991	12 ± 4 days longer
Precipitation	1900–1994	Generally increasing outside tropics, decreasing in Sahel
Heavy precipitation	1910–1990[a]	Growing in importance
Antarctic snowfall	Recent decades	5–20% increase
Extratropical cyclones	Recent decades	Sharp increase in the Pacific sector
Tropical cylones	1945–1993	Decrease in mean annual maximum wind speeds in the Atlantic sector
	1944–1993	Decrease in frequency and intensity in the Atlantic sector
Global mean sea level	20th century	1.8 ± 0.1 mm/year

[a] Shorter records in some regions.

warming of the climate. Sea level has been increasing during the 20th century at a greater rate than during the past 1000 years or more, although the historical record does not show an acceleration within the 20th century.

Questions for further thought

1. Why can't past tropospheric ozone concentrations (prior to the time when direct measurements began) be determined in the same way as for CO_2 or CH_4?
2. How have changes in the atmospheric CO and OH concentrations, since the industrial revolution and during recent decades, been determined?
3. Categorize the reasons why corrections or adjustments are required to the temperature trends deduced from the observed variation of (a) sea surface temperature, (b) land-based air temperature, (c) balloon-based tropospheric temperature, and (d) satellite-based tropospheric temperature.
4. Why are measured long-term trends in the atmospheric water vapour content less reliable than the measured interannual variability in the amount of water vapour? In what region of the atmosphere are trends least reliable? Why is this region particularly significant?

Climatic change: from emissions to climate system response

In Part I we presented an overview of the climate system, compared the major natural and anthropogenic causes of climatic change, and identified the major human sources of GHGs and of aerosols or aerosol precursors. We also examined the changes in the composition of the Earth's atmosphere and in temperature that have occurred over the past century or so, but without, at this stage, attempting to explain past temperature changes.

In Part II we analyse and explain the sequence of processes involved in nature in going from emissions to climatic change and sea level rise. To do this, we first introduce the tools that are used to project changes in the atmospheric composition, in climate, and in sea level: a hierarchy of mathematical models of the climate system and of the atmospheric and oceanic components of the climate system. This is the subject of Chapter 5. In Chapter 6, we lay out the basic principles involved in the climatic response to perturbations in the radiative energy balance. Chapters 7–10 deal with the sequence of processes involved in projecting future climatic change and sea level rise. The first set of processes, discussed in Chapter 7, involves the global carbon cycle and its response both to increasing atmospheric CO_2 concentration and to the induced changes in temperature. This translates emission scenarios into concentration scenarios. We also discuss, within this chapter, the much simpler set of processes involved in translating emissions of other greenhouse gases into concentration scenarios. Next, in Chapter 8, the characteristics of the climate associated with a doubling of the pre-industrial CO_2 concentration, as simulated by three-dimensional atmospheric models coupled to an ocean surface layer, are discussed. We examine simulated changes in regional temperature and soil moisture on a seasonal basis, in tropical and mid-latitude storm systems, and in such large-scale phenomena as the Asian monsoon and El Niño. In Chapter 9, we discuss the time-dependent or "transient" climatic response to changes in radiative forcing as simulated using coupled atmosphere–ocean models. Model projections of transient climatic change resulting from increases in GHG concentrations, with and without increases in aerosols, and with and without various natural forcing mechanisms, are compared with the observed climatic change over the 20th century. Chapter 10 closes Part II with a discussion of sea level rise.

Chapter 5

Models used in projecting future climatic change and sea level rise

In order to project the impact of human perturbations on the climate system, it is necessary to calculate the effects of all the key processes operating in the climate system. These processes can be represented in mathematical terms, but the complexity of the system means that the calculations can be performed in practice only on a computer. The mathematical formulation is therefore implemented in a computer program, which is referred to as a "model". If the model includes enough of the components of the climate system to be useful for simulating the climate (at a minimum, the atmosphere and ocean), it is commonly called a "climate model". A climate model that explicitly included all our current understanding of the climate system would be too complex to run on any existing computer. Instead, simplifications are made to a varying extent. The more that nature is simplified in constructing a model, the faster the model can be run or the less powerful the computer that is needed because fewer calculations are performed. In this chapter the major types of models and the simplifications used to create them are outlined for each of the major steps involved in simulating the climate and sea level response to anthro-pogenic emissions of GHGs and aerosols. The strengths and weaknesses of the major types of models are also outlined in this chapter. The discussion found in this chapter draws heavily upon Harvey *et al.* (1997).

5.1 A hierarchy of atmosphere and ocean climate models

The most detailed model of a particular process is one which is based on funda-mental physical principles which are believed to be invariant. Such a model would be applicable to any climate at all. Examples of such principles include Newton's laws of motion, conservation of energy, conservation of mass, the ideal gas law, and the radiative properties of different gases. In order to represent the process in a way which can be used in a climate model, additional, simplifying, assumptions have to be introduced. In some cases, empirically derived relationships are included. However, because such empirical relationships are not rigorously derived from fundamental principles, the observed correlations might not be applicable if the underlying conditions change – that is, as the climate changes.

In a model, physical quantities which vary continuously in three dimen-sions are represented by their values at a finite number of points arranged in a three-dimensional grid (in the case of 3-D models). This is necessary because only a finite number of calculations can be performed. The spacing between the points of the grid is the "spatial resolution". The finer the resolution, the more points, and the more calculations that need to be done. Hence, the resolution is limited by the computing resources available. The typical resolution that can be used in a climate model is hundreds of kilometres in the horizontal. Many important elements of the climate system (e.g., clouds, land surface variations) have scales much smaller than this. Detailed models at high resolution are available for such processes by themselves, but these are computationally too expensive to be included in a climate model, and the climate model has to represent the effect of these sub-grid-scale processes on the climate system at its coarse grid-scale. A formulation of the effect of a small-scale process on the large scale is called a *parameterization*. All climate models use parameterization to some extent.

Another kind of simplification used in climate models is to average over a complete spatial dimension. Instead of, for instance, a three-dimensional longitude–latitude–height grid, one might use a two-dimensional latitude–height grid, with each point being an average over all longitudes at its latitude and height. When the dimensionality is reduced, more processes have to be parameterized but less computer time is required.

Some of the main types of models for the atmospheric and oceanic components of the climate system are as follows:

One-dimensional radiative–convective atmospheric models

These models are globally (horizontally) averaged but contain many layers within the atmosphere. They treat processes related to the transfer of solar and infrared radiation within the atmosphere in considerable detail, and are particularly useful for computing the radiative perturbation associated with changes in the atmosphere's composition (e.g. Lal and Ramanathan, 1984; Ko *et al.*, 1993).

One-dimensional upwelling–diffusion ocean models

In this model, which was first applied to questions of climatic change by Hoffert *et al.* (1980), the atmosphere is treated as a single well-mixed box that exchanges heat with the underlying ocean and land surface. The ocean is treated as a 1-D column that represents a horizontal average over the real ocean, excluding the limited regions where deep water forms and sinks to the ocean bottom, which are treated separately. Figure 5.1 illustrates this model. The sinking in polar regions is represented by the pipe to the side of the column. This sinking and the compensating upwelling within the column represent the global-scale advective overturning (the thermohaline circulation) (see Section 2.2).

One-dimensional energy balance models

In these models the only dimension that is represented is the variation with latitude; the atmosphere is averaged vertically and in the east–west direction, and often combined with the surface to form a single layer. The multiple

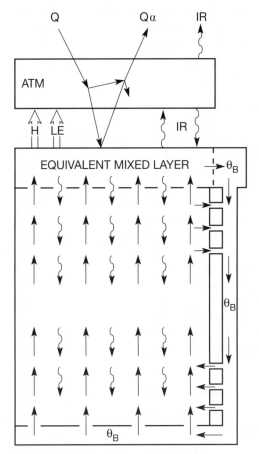

Figure 5.1 Illustration of the one-dimensional upwelling–diffusion model. The model consists of boxes representing the global atmosphere and ocean mixed layer, underlain by a deep ocean. Bottom water forms at temperature Θ_B, sinks to the bottom of the ocean, and upwells. The temperature profile in the deep ocean is governed by the balance between the upwelling of cold water and downward diffusion of heat from the mixed layer. Latent (LE) and sensible (H) heat, and infrared radiation (IR) are transferred between the atmosphere and the surface. Q is solar radiation. Redrafted from Harvey and Schmeider (1985).

processes of north–south heat transport by the atmosphere and oceans are usually represented as diffusion. These models have provided a number of useful insights concerning the interaction of horizontal heat transport feedbacks and high-latitude feedbacks involving ice and snow (e.g. Held and Suarez, 1974).

Two-dimensional atmosphere and ocean models
Several different two-dimensional (latitude–height or latitude–depth) models of the atmosphere and oceans have been developed. The two-dimensional models permit a more physically based computation of horizontal

heat transport than in one-dimensional energy balance models. In some two-dimensional ocean models (e.g., Wright and Stocker, 1991) the intensity of the thermohaline overturning is determined by the model itself, while in others (e.g., de Wolde *et al.*, 1995) it is prescribed, as in the one-dimensional upwelling–diffusion model.

Three-dimensional atmosphere and ocean general circulation models

The most complex atmosphere and ocean models are the three-dimensional atmospheric general circulation models (AGCMs) and ocean general circulation models (OGCMs), both of which are extensively reviewed in Gates *et al.* (1996). Table 5.1 lists some of the AGCM and/or AOGCM modelling groups. These models divide the atmosphere or ocean into a horizontal grid with a typical resolution of 2–4° latitude by 2–4° longitude in the latest models, and typically 10–20 layers in the vertical. They directly simulate winds, ocean currents, and many other features and processes of the atmosphere and oceans. Figure 5.2 provides a schematic illustration of the major processes that occur within a single horizontal grid cell of an AOGCM. Coupled atmosphere and ocean GCMs (AOGCMs) automatically compute the feedback processes associated with water vapour, clouds, and seasonal snow and ice. They also compute the uptake of heat by the oceans, which delays and distorts the surface temperature response but contributes to sea level rise through expansion of ocean water as it warms.

Table 5.1 Acronyms for most of the AGCM and AOGCM modelling groups

Acronym	Definition
ARPEGE[a]	Action de Recherche Petite Echelle Grande Echelle, Metéo France, Toulouse, France
BMRC	Bureau of Meteorology Research Centre, Melbourne, Australia
CCC	Canadian Climate Centre, Victoria, British Columbia
CSIRO	Commonwealth Scientific and Industrial Research Organization, Aspendale, Australia
GFDL	Geophysical Fluid Dynamics Laboratory, Princeton, New Jersey
GISS	Goddard Institute for Space Studies, New York
GLA	Goddard Laboratory for Atmospheres, Greenbelt, Maryland
LMD	Laboratoire de Météorologie Dynamique, Paris
LODYC	Laboratoire d'Océanographie Dynamique et de Climatologie, Paris
MPI	Max-Planck-Institut für Meteorologie, Hamburg, Germany
NCAR	National Center for Atmospheric Research, Boulder, Colorado
NRL	Navy Research Laboratory, Monterey, California
UKMO	United Kingdom Meteorological Office, Bracknell, UK (now the Hadley Centre for Climate Prediction and Research)

[a] This acronym pertains to the name of a model, while all the other acronyms pertain to the name of a research centre or centres.

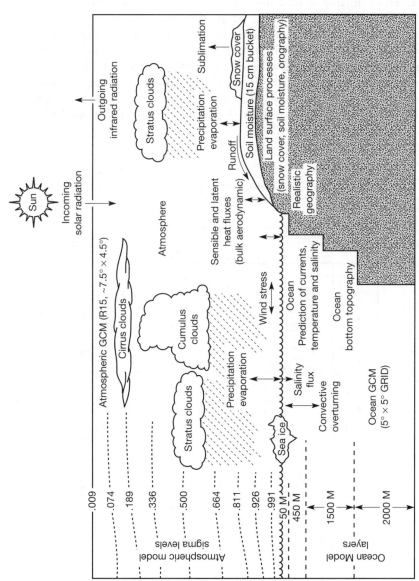

Figure 5.2 Illustration of the major processes occurring inside a typical horizontal grid cell of a coupled atmosphere-ocean general circulation model (AOGCM). Reproduced from Washington and Meehl (1989). Many current models have twice as fine a horizontal resolution as shown here.

5.2 Models of the carbon cycle

The upwelling–diffusion model that was described in Section 5.1 can be used to model the oceanic part of the carbon cycle, as in the work of Hoffert *et al.* (1981) and Jain *et al.* (1995). The global mean atmosphere–ocean exchange of CO_2, the vertical mixing of total dissolved carbon by thermohaline overturning and diffusion, and the sinking of particulate material produced by biological activity can all be represented in this model. An extension of the 1-D upwelling–diffusion model is presented in Figure 5.3, in which separate northern and southern hemisphere regions are explicitly included. The properties of water that upwells as part of the thermohaline circulation – often referred to as "bottom water" – are determined through air–sea exchange and convective mixing in the polar regions. A two-dimensional ocean model has

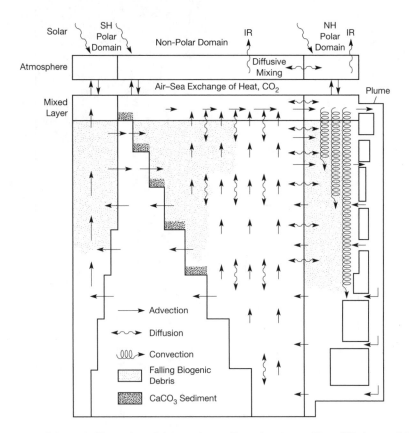

Figure 5.3 Schematic illustration of the quasi-one-dimensional, upwelling-diffusion model of Harvey and Huang (1999), showing the processes of diffusion, advection, convective mixing, and the biological pump. Unlike the 1-D upwelling model shown in Figure 5.1, the polar sea regions where bottom water forms or where combined upwelling and convective mixing occurs are explicitly represented. The subsurface SH polar domain is drawn separated from the non-polar domain for clarity only.

also been used as the oceanic component of the global carbon cycle (Stocker *et al.*, 1994). Finally, OGCMs can be used as the oceanic component of the global carbon cycle, in which the model-computed ocean currents and other mixing processes are used, in combination with simple representations of biological processes and air–sea exchange (e.g., Bacastow and Maier-Reimer, 1990; Najjar *et al.*, 1992).

The terrestrial biosphere can be represented by a series of interconnected boxes, where the boxes represent components such as leafy material, woody material, roots, detritus, and one or more pools of soil carbon. Each box can be globally aggregated such that, for example, the detrital box represents all the surface detritus in the world. The commonly used, globally aggregated box models are quantitatively compared in Harvey (1989). Figure 5.4 gives an example of a globally aggregated terrestrial biosphere model, where numbers inside boxes represent the steady-state amounts of carbon (Gt) prior to human disturbances, and the numbers between boxes represent the annual rates of carbon transfer (Gt yr^{-1}).

The role of the terrestrial biosphere in global climatic change has also been simulated using relatively simple models of vegetation on a global grid with resolution as fine as 0.5° latitude × 0.5° longitude. These models have been

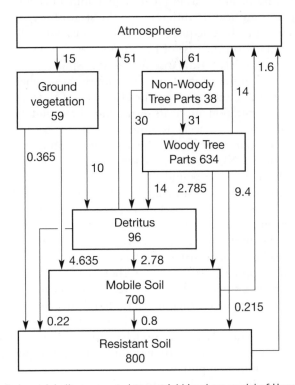

Figure 5.4 The six-box, globally aggregated terrestrial biosphere model of Harvey (1989). The numbers in each box are the steady-state amounts of carbon (in Gt), while the arrows between the boxes represent the annual transfers.

used to evaluate the impact on net ecosystem productivity of higher atmospheric CO_2 (which tends to stimulate photosynthesis and improve the efficiency of water use by plants) and warmer temperatures (which can increase or decrease photosynthesis and increase decay processes). Grid point models of the terrestrial biosphere have also been used to assess the effect on the net biosphere–atmosphere CO_2 flux of hypothetical (or GCM-generated) changes in temperature and/or atmospheric CO_2 concentration, but generally without allowing for shifts in the ecosystem type at a given grid point as climate changes. More advanced ecosystem models are being developed and tested that link biome models (which predict changing ecosystem types) and ecophysiological models (which predict carbon fluxes) (e.g. Plöchl and Cramer, 1995).

Rather detailed models of the marine biosphere, involving a number of species and interactions, have also been developed and applied to specific sites or regions (e.g., Gregg and Walsh, 1992; Sarmiento et al., 1993; Antoine and Morel, 1995).

5.3 Models of atmospheric chemistry and aerosols

As discussed in Section 2.8, atmospheric chemistry is central to the distribution and amount of ozone in the atmosphere. The dominant chemical reactions and sensitivities are significantly different for the stratosphere and troposphere. These processes can be adequately modelled only with three-dimensional atmospheric models (in the case of the troposphere) or with two-dimensional (latitude–height) models (in the case of the stratosphere). Atmospheric chemistry is also critical to the removal of CH_4 from the atmosphere and, to a lesser extent, all other greenhouse gases except H_2O and CO_2. In the case of CH_4, a change in its concentration affects its own removal rate and, hence, subsequent concentration changes. An accurate simulation of changes in the removal rate of CH_4 requires specification of the concurrent concentrations of other reactive species, in particular NO_x (nitrogen oxides), CO (carbon monoxide) and the VOCs (volatile organic compounds), and use of a model with latitudinal and vertical resolution. However, simple, globally averaged models of chemistry–climate interactions have been developed. These models treat the global CH_4–CO–OH cycle in a manner which takes into account the effects of the heterogeneity of the chemical and transport processes, and provide estimates of future global or hemispheric mean changes in the chemistry of the Earth's atmosphere. An even simpler approach, adopted by Osborn and Wigley (1994), is to treat the atmosphere as a single well-mixed box but to account for the effects of atmospheric chemistry by making the CH_4 lifetime depend on CH_4 concentration in a way that roughly mimics the behaviour of the above-mentioned globally averaged models or of models with explicit spatial resolution. Atmospheric O_3 and CH_4 chemistry has not yet been incorporated in AGCMs used for climate simulation purposes.

Atmospheric chemistry is also central to the distribution and radiative properties of aerosols, although chemistry is only part of what is required in order to simulate the effects of aerosols on climate. The key processes that need to be represented are the source emissions of aerosols or aerosol precursors; atmospheric transport, mixing, and chemical and physical transformation; and removal processes (primarily deposition in rainwater and direct dry deposition onto the Earth's surface). Since part of the effect of aerosols on climate arises because they serve as cloud condensation nuclei, it is also important to be able to represent the relationship between changes in the aerosol mass input to the atmosphere and, ultimately, the radiative properties of clouds. Establishing the link between aerosol emissions and cloud properties, however, involves several poorly understood steps and is highly uncertain.

5.4 Models of ice sheets

High resolution (20 km × 20 km horizontal grid), two- and three-dimensional models of the polar ice sheets have been developed and used to assess the impact on global mean sea level of various idealized scenarios for temperature and precipitation changes over the ice sheets (e.g., Huybrechts and Oerlemans, 1990; Huybrechts et al., 1991). AGCM output has also recently been used to drive a three-dimensional model of the East Antarctic ice sheet (Verbitsky and Saltzman, 1995), but has not yet been used to assess the possible contribution of changes in mountain glaciers to future sea level rise. Output from high-resolution ice sheet models can be used to develop simple relationships in which the contribution of ice sheet changes to future sea level is scaled with changes in global mean temperature.

5.5 Utilization of simple and complex models

As shown in the preceding discussion, models of the climate system range from relatively simple, schematic representations of the major processes, to quite complex models that attempt to represent climate processes in as much detail and with as high a resolution as computer limitations will permit. Both simple and complex models have important but different roles to play in projecting future climatic change due to human activities.

Simple models allow investigation of the basic relationships between the components of the climate system, the factors driving climatic change, and the overall response of the system. Because simple models represent only the most critical processes, they are relatively easy to understand and inexpensive to run, so that multiple diagnostic tests can be executed. They are useful mainly for exploring global-scale questions. Three-dimensional models, on the other hand, provide the only prospect for reliably simulating the key processes that determine climate sensitivity and the longer-term feedbacks involving the terrestrial and marine biosphere. Results from such models provide one means of determining the values of the aggregate feedback parameters that are used in simpler models. Complex models are also needed for the simulation of regional climatic change and of variability on short time scales. On the other hand, complex models are computationally costly, are sometimes difficult to understand, and require high-resolution data inputs, which in some cases simply do not exist. They produce outputs which contain substantial temporal and spatial variability (sometime referred to as "noise"); this makes analysis of their results a complicated task, as is the case for the real climate system.

The subcomponents of simple models can be constrained to replicate the overall behaviour of the more complex model subcomponents. For example, the climate sensitivity of simple models can be made to equal that of any particular AGCM by altering a single model parameter – the coefficient governing infrared emission to space – whose value implicitly accounts for the net, global mean effect of all the fast feedback processes (Section 6.5) that influence infrared radiation. Similarly, the vertical diffusion coefficient and the upwelling velocity can be readily altered such that the oceanic uptake of heat (and associated sea level rise) closely matches that of any given OGCM. Globally aggregated biosphere models can be adjusted to replicate the sensitivity to atmospheric CO_2 and temperature changes obtained by regionally distributed models. This allows the simple models to span the range of behaviour of the more detailed, regionally resolved model components, which in turn makes them ideal tools for examining interactions between the model components and for the analysis of alternative emission scenarios.

Questions for further thought

1. What is the difference between (a) a fundamental physical law, and (b) a parameterization?
2. Why are parameterizations required?
3. Give examples of fundamental physical laws used in 3-D atmospheric general circulation models (AGCMs), and of places where parameterizations are required in AGCMs.

Chapter 6

The physics of the greenhouse effect, radiative forcing, and climate sensitivity

As noted in Section 2.2, the term "greenhouse effect" refers to the reduction in outgoing infrared radiation to space due to the presence of the atmosphere. It can be and has been directly measured, based on the surface emission (which depends only on temperature) and the emission at the top of the atmosphere as observed by satellites. In this chapter the underlying physics of this effect is explained in very simple terms. This is followed by a discussion of the heating perturbation that occurs when GHG or aerosol concentrations change – the so-called "radiative forcing". The concept of fast and slow feedbacks and of climate sensitivity is then introduced, followed by a discussion of the relationship between radiative forcing and climate sensitivity for a variety of important radiative forcing mechanisms.

6.1 Physics of the greenhouse effect

All objects above absolute zero (0 K or −273°C) emit electromagnetic radiation. For objects at typical earth-atmosphere temperatures, the emitted radiation falls almost entirely in the infrared (IR) part of the spectrum (wavelengths of 4.0 to 50.0 μm). The maximum amount of radiation that can be emitted is given by

$$F_{max} = \sigma T^4 \tag{6.1}$$

where $\sigma = 5.664 \times 10^{-8}\,\text{W m}^{-2}\,\text{K}^{-4}$ and is called the Stefan–Boltzmann constant, and T is the absolute temperature in telvins. Equation (6.1) is known as the *Stefan–Boltzmann Law*. Objects that emit the maximum amount of radiation are called *blackbodies*. The ratio of actual emission to blackbody emission is called the *emissivity*, ε. Actual emission is thus given by

$$F = \varepsilon \sigma T^4 \tag{6.2}$$

The land and ocean surface and clouds thicker than cirrus clouds (thin, wispy clouds) emit almost as blackbodies ($\varepsilon \approx 1.0$). The clear-sky atmosphere has an emissivity of 0.4–0.8, while cirrus clouds have a typical emissivity of 0.2. The Earth, atmosphere, and clouds also absorb IR radiation. The ratio of absorbed radiation to incident radiation is called the absorptivity. Kirchhoff's Law states that absorptivity is equal to emissivity, which implies that a blackbody will absorb all of the incident infrared radiation, while the atmosphere will absorb only part of the incident radiation, the balance being transmitted. Reflection or scattering of IR radiation is generally negligible and so will not be considered here.

With this background, we can analyse the greenhouse effect with the aid of the simple two-box model shown in Figure 6.1. The lower box represents the Earth's surface, which has temperature T_s and emits σT_s^4 (since $\varepsilon_s = 1.0$). The upper box represents the atmosphere, which has temperature T_a and emissivity ε_a. It absorbs $\varepsilon_a \sigma T_s^4$ from below, allows $(1 - \varepsilon_a)\sigma T_s^4$ to be transmitted to space, and re-emits an amount $\varepsilon_a \sigma T_a^4$ to space. The total emission to space is given by

$$F_{space} = (1 - \varepsilon_a)\sigma T_s^4 + \varepsilon_a \sigma T_a^4 \qquad (6.3)$$

which can be rewritten as

$$F_{space} = \sigma T_s^4 - \varepsilon_a (\sigma T_s^4 - \sigma T_a^4) \qquad (6.4)$$

Equation (6.4) represents the emission to space as the sum of the surface emission and a term that depends on the atmosphere. Since, in general, $T_a < T_s$, the second term (with the negative sign) is negative, that is, it subtracts from the surface emission, thereby reducing the emission to space. This term is properly called the "atmospheric effect", but corresponds to what is popularly known as the "greenhouse effect". In physical terms, what is happening is that the atmosphere absorbs a portion of the radiation stream from the surface ($\varepsilon_a \sigma T_s^4$) and replaces this stream with its own emission ($\varepsilon_a \sigma T_a^4$). As long as the atmosphere is colder than the surface, the replacement stream is less than the original stream, so that the net emission of infrared radiation to space is reduced. Equation (6.4) makes it clear that the larger the atmospheric emissivity, the greater the reduction in outgoing radiation to space (even though the contribution from the atmosphere itself increases) and the stronger the greenhouse effect.

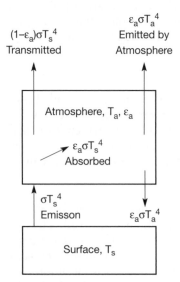

Figure 6.1 A two-box model of the atmosphere–surface system, showing the transfer of IR radiation. T_a and T_s are the atmospheric and surface temperatures, respectively, and ε_a is the atmospheric emissivity.

The real atmosphere does not consist of a layer with a single temperature. Rather, there is a variation of temperature with height. T_a can be regarded as an *effective* atmospheric temperature, which depends on the temperature of each layer weighted by the fraction of total emissivity in that layer, and summed over all layers.

The primary gases that contribute to the atmospheric emissivity are H_2O, CO_2, O_3, and CH_4. Carbon dioxide and CH_4 are fairly evenly mixed throughout the atmosphere. The ozone concentration varies primarily with height, but this variation is relatively fixed in time. The water vapour concentration, however, varies strongly in space and in time. If a given amount of water vapour were to be distributed higher in the troposphere, this would increase the emissivity of the cold layers and decrease the emissivity of the warm layers, thereby reducing the effective radiating temperature. From Equation (6.4), this would increase the greenhouse effect (the emission to space would be reduced). Another way to view this result is as follows: increasing the emissivity of a warm layer or a cold layer by a given amount would increase the absorption of surface radiation by the same amount, but the increase in emission from the atmosphere would be smaller if the extra emissivity is added to the colder layer. Thus, water vapour (or any other GHG) is more effective as a greenhouse gas in the upper troposphere (where temperatures are colder) than in the lower troposphere.

Although the atmosphere emits radiation to the surface, this downward emission does not constitute the greenhouse effect. The trapping of outgoing radiation is far more important to the surface temperature than is the downward IR emission. This is because the atmosphere and surface are tightly coupled by sensible and latent heat fluxes. Downward emission redistributes heat between the surface and atmosphere without affecting the total amount of heat available, whereas the partial trapping of radiation emitted to space alters the heat balance of the combined atmosphere + surface system. This point can be demonstrated quantitatively (albeit approximately) by modifying the two-box model to include latent and sensible heat fluxes between the surface and atmosphere, as shown in Box 6.1.

Box 6.1 Further analysis of the natural greenhouse effect

In this box it will be demonstrated that the effect on surface temperature due to the downward emission of infrared radiation by greenhouse gases is an order of magnitude smaller than the effect on surface temperature arising from the reduction in outgoing emission to space.

Consider a two-box model, where Q_a^* and Q_s^* are the rate of absorption of solar energy by the atmosphere and surface, respectively, $L\uparrow$ and $L\downarrow$ are the upward and downward infrared fluxes at the surface, and H is the sum of the sensible and latent heat transfer between the surface and atmosphere. The emission to space, L_{sp}, is equal

(continued)

(continued)

to $L\uparrow - G$, where G is the greenhouse effect. For a steady-state climate, the net energy flow into each box is zero. Thus, we can write

$$Q_a^* + G + H - L\downarrow = 0 \tag{6.1.1}$$

and

$$Q_s^* + L\downarrow - L\uparrow - H = 0 \tag{6.1.2}$$

If a heating perturbation is applied, the temperatures of both the surface and atmosphere, and the associated energy flows, will adjust in such a way as to restore zero net energy input for both boxes. If $F(T_0)$ is the value of energy flux F for the initial steady-state climate, characterized by temperature T_0, then the value of F for some new temperature T can be approximated by the first-order Taylor Series expansion, $F(T) = F(T_0) + (dF/dT)\Delta T$. Now suppose that we are going to change $L\downarrow$ by an amount $\Delta L\downarrow$. For simplicity, we shall neglect changes in the absorption of solar energy $(Q_a^* + Q_s^*)$, as such changes do not significantly change the results. The equations expressing the new balance that will eventually be achieved by the atmosphere and surface are

$$-\Delta L\downarrow + \left(\frac{\partial G}{\partial T_a} + \frac{\partial H}{\partial T_a} - \frac{\partial L\downarrow}{\partial T_a}\right)\Delta T_a + \left(\frac{\partial G}{\partial T_s} + \frac{\partial H}{\partial T_s}\right)\Delta T_s = 0 \tag{6.1.3}$$

and

$$\Delta L\downarrow + \left(\frac{\partial L\downarrow}{\partial T_a} - \frac{\partial H}{\partial T_a}\right)\Delta T_a - \left(\frac{\partial L\uparrow}{\partial T_s} + \frac{\partial H}{\partial T_s}\right)\Delta T_s = 0 \tag{6.1.4}$$

respectively. Equations (6.1.3) and (6.1.4) can be added to yield

$$\left(\frac{\partial G}{\partial T_a}\right)\Delta T_a + \left(\frac{\partial G}{\partial T_s} - \frac{\partial L\uparrow}{\partial T_s}\right)\Delta T_s = 0 \tag{6.1.5}$$

As discussed in Section 6.1, G is given by $\varepsilon_a(\sigma T_s^4 - \sigma T_a^4)$, where T_a is the effective radiating temperature of the atmosphere. The atmospheric emissivity, ε_a, depends on the concentration of gases such as CO_2 and CH_4, which we can regard as being independent of temperature, and on the concentration of water vapour. As temperature increases, the saturation vapour pressure and hence the amount of water vapour in the atmosphere increase rapidly, but as the water vapour amount increases, the emissivity saturates at a maximum value of 1.0. For temperature deviations around some reference temperature T_0, it is convenient to approximate ε_a as

$$\varepsilon_a = a + b(T_a - T_0)^n \tag{6.1.6}$$

(continued)

(continued)

Thus, the greenhouse effect G is given by

$$G = (a + b(T_a - T_0)^n)(\sigma T_s^4 - \sigma T_a^4)$$

The total turbulent flux H can be written as $H = c(T_s - T_a)$, while $L\!\uparrow = \sigma T_s^4$ and $L\!\downarrow = \varepsilon_a(T_a + \delta T)^4$, where δT accounts for the fact that the effective temperature for downward radiation is greater than for upward radiation. Given a present-day atmospheric emissivity, ε_0, of about 0.73 and choosing $n = 1$ and $T_0 = 200$ K, we require that $a = 0.4086$ and $b = 0.00824$. Appropriate values of the other constants appearing in the above expressions are as follows: $\sigma = 5.67 \times 10^{-8}$ W m^{-2} K^{-4}, $T_s = 289$ K, $T_a = 239$ K, $\delta T = 30$ K, and $c = 20$ W m^{-2} K^{-4}. We can then evaluate the partial derivatives appearing in Equations (6.1.3)–(6.1.5) as follows:

$$\frac{\partial G}{\partial T_a} = \frac{d\varepsilon_a}{\partial T_a}(\sigma T_s^4 - \sigma T_a^4) - \varepsilon_0 4\sigma T_a^3 = 1.734 - 2.260$$

$$= -0.526 \text{ W m}^{-2}\text{K}^{-1} \qquad (6.1.7)$$

$$\frac{\partial G}{\partial T_s} = \varepsilon_0 4\sigma T_s^3 = 4.00 \text{ W m}^{-2}\text{K}^{-1} \qquad (6.1.8)$$

$$\frac{\partial L\!\uparrow}{\partial T_s} = 4\sigma T_s^3 = 5.474 \text{ W m}^{-2} \text{ K}^{-1} \qquad (6.1.9)$$

$$\frac{\partial L\!\downarrow}{\partial T_a} = 4\varepsilon_0\sigma(T_a + \delta T)^3 = 3.22 \text{ W m}^{-2}\text{K}^{-1} \qquad (6.1.10)$$

$$\frac{\partial H}{\partial T_s} = -\frac{dH}{\partial T_a} = c = 20 \text{ W m}^{-2}\text{K}^{-1} \qquad (6.1.11)$$

The first term on the right-hand side of Equation (6.1.7) is the effect of increasing atmospheric emissivity as atmospheric temperature increases, and it is positive. The second term is the effect of increasing atmospheric temperature with fixed emissivity and surface temperature, and it is negative. That is, the greenhouse effect decreases as atmospheric temperature increases with fixed surface temperature and emissivity.

Inserting the appropriate derivative values into Equation (6.1.5) leads to the result that $\Delta T_a = -2.80\Delta T_s$. From Equation (6.1.4) we obtain

$$\Delta T_s = \frac{\Delta L\!\downarrow}{2.80\left(\dfrac{\partial L\!\downarrow}{\partial T_a} - \dfrac{\partial H}{\partial T_a}\right) + \left(\dfrac{\partial L\!\uparrow}{\partial T_s} + \dfrac{\partial H}{\partial T_s}\right)} = \frac{\Delta L\!\downarrow}{90.5} \qquad (6.1.12)$$

(continued)

(continued)

and from the above, it follows that

$$\Delta T_a = -\frac{\Delta L\!\downarrow}{32.3} \tag{6.1.13}$$

Thus, an increase in $L\!\downarrow$ by $10\,\mathrm{W\,m^{-2}}$ would cool the atmosphere by about $0.3\,\mathrm{K}$ and warm the surface by about $0.1\,\mathrm{K}$. The reason the temperature responses are so low is the strong coupling between the atmosphere and surface, and the most important contributor to this is the turbulent heat flux. This is reflected in the fact that the largest terms in the denominator to Equation (6.1.12) are the $\partial H/\partial T_a$ and $\partial H/\partial T_s$ terms.

Now, assume that instead of perturbing $L\!\downarrow$, we perturb G by an amount ΔG. As before, T_a and T_s will adjust so as to restore zero net energy balance for both the atmosphere and surface. The equations expressing the new balance are

$$\Delta G + \left(\frac{\partial G}{\partial T_a} + \frac{\partial H}{\partial T_a} - \frac{\partial L\!\downarrow}{\partial T_a}\right)\Delta T_a + \left(\frac{\partial G}{\partial T_s} + \frac{\partial H}{\partial T_s}\right)\Delta T_s = 0 \tag{6.1.14}$$

and

$$\left(\frac{\partial L\!\downarrow}{\partial T_a} - \frac{\partial H}{\partial T_a}\right)\Delta T_a - \left(\frac{\partial L\!\uparrow}{\partial T_s} + \frac{\partial H}{\partial T_s}\right)\Delta T_s = 0 \tag{6.1.15}$$

From Equation (6.1.15) it follows that $\Delta T_a = 1.1\Delta T_s$. Substituting into Equation (6.1.14) yields

$$\Delta T_s = -\frac{\Delta G}{1.1\left(\dfrac{\partial G}{\partial T_a} + \dfrac{\partial H}{\partial T_a} - \dfrac{\partial L\!\downarrow}{\partial T_a}\right) + \left(\dfrac{\partial G}{\partial T_s} + \dfrac{\partial H}{\partial T_s}\right)} = \frac{\Delta G}{2.12} \tag{6.1.16}$$

and, from the above

$$\Delta T_a = \frac{\Delta G}{1.93} \tag{6.1.17}$$

Thus, an externally imposed $10\,\mathrm{W\,m^{-2}}$ increase in G requires a warming of about $5\,\mathrm{K}$. Careful inspection of Equation (6.1.16) indicates that the two turbulent heat flux terms, $\partial H/\partial T_a$ and $\partial H/\partial T_s$, largely cancel, so that the temperature response is governed by the radiative damping (which is much weaker). Furthermore, the total radiative damping appearing in Equation (6.1.16) is related, and very similar in magnitude, to the total radiative damping to space, $\partial L_{sp}/\partial T_a$, as the reader should be able to show.

(continued)

(continued)

To summarize, when the downward infrared flux is altered, there are opposing radiative perturbations for the atmosphere and surface, and the response of both is governed by turbulent heat exchange. When a net heating perturbation is applied to the atmosphere + surface system, the response is governed by the radiative damping to space. Since radiative damping to space is much weaker than turbulent heat flux damping, the surface (or air) temperature response is much larger in the second case than in the first case.

6.2 Radiative forcing

Changes in surface temperature are driven by changes in the net radiation *at the tropopause*, not at the surface or at the top of the atmosphere. This change in net radiation is called the *radiative forcing*, because it is what forces or drives the change in surface climate. The radiative forcing involves the change in the net downward flux of solar energy at the tropopause, the change in the upward emission of infrared radiation at the tropopause, and the change in the downward emission of infrared radiation from the stratosphere. The change in net radiation at the tropopause before any temperatures are allowed to change is called the *instantaneous radiative forcing*. As stratospheric temperatures change in response to the perturbation in the stratospheric radiative energy balance, the downward emission of infrared radiation will change, which will alter the subsequent change in the tropospheric and surface temperatures. Because the stratosphere responds quickly (within months) and independently of the surface–troposphere system, the effect of any changes in stratospheric temperatures should be included when computing the radiative forcing. This gives the *adjusted radiative forcing*.

Figure 6.2 shows the change in global mean fluxes for a CO_2 doubling, before and after adjustment of stratospheric temperatures. The global mean instantaneous forcing, as computed by a variety of researchers, is about 4.0–4.5 W m^{-2} (Cess *et al.*, 1993). Of this, about 1.5 W m^{-2} – or 40% – is due to extra downward emission from the stratosphere. This extra downward radiation causes the stratosphere to cool at the same time that the surface and troposphere warm. The cooling of the stratosphere in turn reduces the downward emission of radiation from the stratosphere by about 0.5 W m^{-2} (for a doubling of CO_2), thereby offsetting part of the initial increase in downward emission caused by the increase in CO_2. Hence, the global mean adjusted radiative forcing for a CO_2 doubling is 3.5–4.0 W m^{-2}. Table 6.1 lists the instantaneous and adjusted direct radiative forcings for changes in the concentration of selected GHGs. The instantaneous and adjusted radiative forcings are almost identical in all cases except for CO_2, where the adjusted forcing is

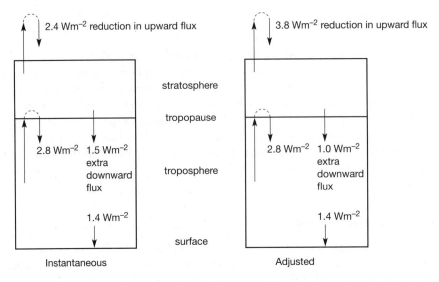

Figure 6.2 Summary of the changes in global mean radiative fluxes associated with a doubling of the atmospheric CO_2 concentration, both before ("Instantaneous") and after ("Adjusted") stratospheric temperatures have fully responded to the change in CO_2 concentration. After stratospheric adjustment, the change in net radiation at the top of the atmosphere equals the change (net forcing) at the tropopause, since otherwise further changes in stratospheric temperature would be required.

Table 6.1 Instantaneous and adjusted direct radiative forcings for changes in the concentration of various greenhouse gases, as given by Shine *et al.* (1995), except for SF_6, which is the average of the clear and cloudy sky forcings as computed by Myhre and Stordal (1997). Instantaneous and adjusted radiative forcings for a wide range of other gases can be found in Myhre and Stordal (1997) and Christidis *et al.* (1997). The difference between the instantaneous and adjusted radiative forcing is due to the change in stratospheric temperature (ΔT_{st}) that is induced by the change in gas concentration, which is given in the last column

Forcing mechanism	Instantaneous forcing ($W\,m^{-2}$)	Adjusted forcing ($W\,m^{-2}$)	ΔT_{st}
CO_2: $300 \rightarrow 600$ ppmv	4.63	4.35	Cooling
CH_4: $0.28 \rightarrow 0.56$ ppmv	0.53	0.52	Slight cooling
N_2O: $0.16 \rightarrow 0.32$ ppmv	0.96	0.93	Slight cooling
CFC-11: $0 \rightarrow 1$ ppbv	0.21	0.22	Warming
CFC-12: $0 \rightarrow 1$ ppbv	0.26	0.28	Warming
SF_6: $0 \rightarrow 1$ ppbv	0.56	0.62	Warming
Stratospheric O_3: 50% reduction	0.26	−1.23	Strong cooling
Solar luminosity: 2% increase	4.48	4.58	Warming

about 7% smaller than the instantaneous forcing, and for depletion of strato-spheric O_3, where the two forcings differ in sign due to the strong cooling of the stratosphere caused by loss of ozone.

6.3 Concept of radiative damping

The temperatures of the atmosphere and surface tend to adjust themselves such that there is a balance between the absorption of energy from the sun and the emission of infrared radiation to space. If, for example, there were to be an excess of absorbed solar energy over emitted infrared radiation (as occurs with the addition of GHGs to the atmosphere), temperatures would increase but, in so doing, the emission of infrared radiation to space would increase. This would reduce the initial imbalance, and eventually a new balance would be achieved, but at a new, warmer temperature. The more rapidly infrared emis-sion to space increases with increasing temperature, the less the temperature must increase in order to restore balance.

However, as temperatures increase, other quantities also change, which fur-ther alters the net emission of radiation to space. For example, an increase in the amount of water vapour as temperature increases tends to counteract the effect of warmer temperatures in increasing the emission of infrared radiation to space. As a result, infrared emission would increase more slowly and a greater temperature increase would be required to restore radiative balance. A reduction in the amount of solar energy reflected to space would also tend to offset the effect of increasing infrared emission to space. The resulting rate at which the net emission of radiation to space increases is called the *radiative damping*, λ. The term "damping" arises because it is the increase in net emis-sion to space as temperatures warm that limits or dampens the increase in temperature. Given λ, the final or steady-state temperature change is given by

$$\Delta T = \frac{\Delta R}{\lambda} \tag{6.5}$$

where ΔR is the adjusted radiative forcing. A larger radiative damping results in a smaller temperature increase.

6.4 Temperature response without feedbacks

In the absence of any feedbacks except the increase in temperature itself, the only way that the emission of radiation to space increases is through the direct dependence of emission on temperature through the Stefan–Boltzmann Law (Equation (6.1)). The radiative damping in this case is $3.76\,\mathrm{W\,m^{-2}\,K^{-1}}$. A doub-ling of atmospheric CO_2 causes a global mean radiative surplus ΔR of 3.5–4.0 $\mathrm{W\,m^{-2}}$. The temperature increase required to restore radiative balance in this case would be $\Delta R/\lambda = 1.0\,\mathrm{K}$ (a warming of 1.0°C). However, as noted above, the very change in temperature would cause other atmospheric and surface properties to change, which would lead to further alterations in the energy bal-ance and require further temperature changes through a series of feedback processes. These feedback processes are identified next.

6.5 Climate feedbacks

Feedbacks play a major role in translating a radiative forcing into a change in climate. They do this by altering the radiative damping. Among the feedbacks that have to be considered are the following.

Water vapour amount
In a warmer climate the atmospheric concentration of water vapour will increase. Because water vapour is a greenhouse gas, this is a positive feedback – that is, it amplifies the initial change.

Water vapour distribution
If, as the total water vapour amount increases, the percentage increase is greater at higher altitudes than at lower altitudes, this will result in a greater climate warming than if the water vapour amount increases by the same percentage at all altitudes. This is because the higher the water vapour, the more effective it is as a greenhouse gas, as explained earlier in this chapter. Conversely, if the increase in water vapour occurs preferentially in the lower troposphere, the net radiative damping will be larger and the climate sensitivity smaller.

Clouds
Changes in clouds can serve as a positive or negative feedback on climate. Clouds have a cooling effect by reflecting sunlight; this effect depends on the difference between the cloud and surface albedos, and on the amount of incident solar radiation. The smaller the underlying albedo and the greater the incident solar radiation, the greater the increase in the reflection of solar radiation to space due to clouds. The cooling effect thus depends on where and when the clouds occur. Clouds exert a warming effect by absorbing the infrared radiation emitted by the surface and re-emitting a smaller amount of radiation due to the fact that the cloud top is colder than the surface. Since higher clouds are colder and thus emit less radiation, the warming effect of clouds depends on how high they are. High clouds tend to have a net warming effect on climate, while low clouds have a net cooling effect. The net feedback of changes in clouds depends on a large number of variables, so even the sign of the feedback is uncertain.

Atmospheric temperature structure
Changes in the rate of decrease of tropospheric temperature with height (or *lapse rate*) will alter the relationship between surface temperature and infrared emission to space, thereby exerting a feedback effect on climate. If the lapse rate decreases as climate warms, then the upper troposphere warms faster than the lower troposphere, and by more than if the lapse rate were constant. This in turn means that the infrared emission to space increases faster as the surface temperature increases than for a constant lapse rate, so that less surface temperature warming is required in order to restore radiative balance. Thus, a decrease in lapse rate with warming serves as a negative feedback. Conversely, an increase in lapse rate with warming serves as a positive feedback.

Areal extent of ice and snow

A reduction in the area of sea ice and seasonal snow cover on land as climate warms will reduce the surface reflectivity, thereby tending to produce greater warming (a positive feedback). However, concurrent changes in cloud cover that could be induced by the change in ice or snowcover could significantly alter the net feedback (Randall *et al.*, 1994).

Vegetation

Changes in the distribution of different biomes or in the nature of vegetation within a given biome can also lead to changes in the surface reflectivity, thereby exerting a feedback effect on climatic change.

The carbon cycle

The effect of climate on the terrestrial biosphere and the oceans is likely to alter the sources and sinks of CO_2 and CH_4, leading to changes in their atmospheric concentrations and hence causing a radiative feedback.

Of these feedbacks, those involving water vapour, the lapse rate, and clouds respond essentially instantaneously to climatic change, while those involving sea ice and snow respond within a few years. These feedbacks are therefore referred to as "fast" feedbacks. Some vegetation and carbon cycle process are relevant on a time scale of decades. Other feedbacks, such as dissolution of carbonate sediments in the ocean and enhanced chemical weathering on land (both of which tend to reduce the atmospheric CO_2 concentration), require hundreds to thousands of years to unfold. Feedbacks involving continental ice sheets occur over periods of thousands of years.

6.6 Climate sensitivity

The term "climate sensitivity" refers to the ratio of the steady-state increase in the global and annual mean surface air temperature to the global and annual mean radiative forcing. It has units of $K\,(W\,m^{-2})^{-1}$. It is standard practice to include only the fast feedback processes, including changes in water vapour, in the calculation of climate sensitivity, but to exclude possible induced changes in the concentrations of other GHGs (as well as other slow feedback processes).

The climate sensitivity is often quoted as the globally average warming once the climate has fully adjusted to an imposed doubling of atmospheric CO_2.[1] However, it is more appropriate to restrict the term "climate sensitivity" to the ratio of the global mean temperature response to the global mean radiative forcing, and to refer to the globally averaged warming range as the equilibrium temperature response for a CO_2 doubling rather than calling it the climate sensitivity.

---

[1] This is generally expected to fall between 1.5°C and 4.5°C although, as discussed later, there are reasons to believe that the lower half of this range is more likely.

A wide variety of experiments with AGCMs indicates that:

- the climate sensitivity is approximately constant and therefore independent of the magnitude (and sign) of the forcing, so that the temperature response varies close to linearly with the forcing; and
- the climate sensitivity does not depend on the spatial (latitudinal) pattern of the forcing in most cases, so that it is the same (to within 10–20%) for different forcing mechanisms.

To the extent that these conditions are true, one can simply add up the individual radiative forcings to get the total global mean radiative forcing, and then apply the climate sensitivity obtained for a CO_2 doubling to determine the global mean temperature response. However, there are two important exceptions: the climate sensitivity for increases in absorbing aerosols and for a change in tropospheric O_3 differs notably from that for other forcing mechanisms.

Apart from the question of whether or not the global mean temperature response depends on the latitudinal variation in the forcing (and not just its global mean value), there is the additional question of the extent to which the *local* temperature response depends on the *local* forcing rather than on the global mean forcing. Again, experiments with 1-D energy balance models and 3-D AGCMs indicate that, when the spatial differences in the forcings are not too large (such as between different well-mixed gases or between well-mixed gases and changes in solar luminosity), the variation in the temperature response with latitude is largely independent of the latitudinal structure of the radiative forcing, and depends mainly on the global mean forcing. That is, there is a roughly constant temperature response pattern, and the absolute responses get multiplied everywhere by roughly the same factor when the global mean forcing changes.

This behaviour is demonstrated by simulations of the temperature response to changes in CO_2 and solar luminosity. Figure 6.3 compares the latitudinal variation in surface–troposphere forcing for a doubling of atmospheric CO_2 and a 2% increase in solar luminosity, while Figure 6.4 gives the latitudinal variation in the zonally (east–west) averaged surface-air warming as simulated by Hansen *et al.* (1997a). In both cases the heating perturbation is strongest at low latitudes, with a particularly sharp dropoff with increasing latitude for an increase in solar luminosity. Nevertheless, in both cases the simulated temperature response is smallest at low latitudes and largest at high latitudes, with roughly the same shape to the variation of the temperature response with latitude. This is because the latitudinal structure of the temperature response is largely determined by feedbacks that are driven by overall warming and not by the spatial structure of the heating perturbation. Thus, although the direct heating effect of an increase in solar luminosity is smallest in polar regions, the warming at lower latitudes induces an increase in the north–south heat flux that overwhelms the direct effect of the heating perturbation at high latitudes and is further amplified by feedbacks involving the melting of ice and snow. A very similar response pattern

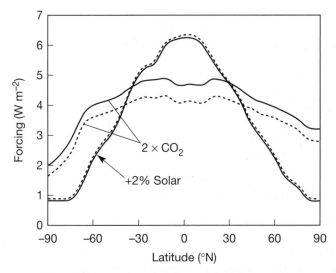

Figure 6.3 Variation with latitude in the zonally and annually averaged instantaneous (solid lines) and adjusted (dashed lines) radiative forcing for a doubling of atmospheric CO_2 and for a 2% increase in solar luminosity, as computed by Hansen *et al.* (1997a). These results were kindly provided in electronic form to facilitate regraphing.

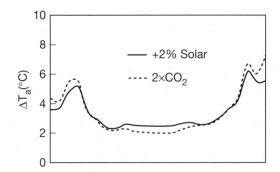

Figure 6.4 Variation with latitude in the zonally and annually averaged surface air warming in response to a doubling of atmospheric CO_2 and a 2% increase in solar luminosity, as computed by Hansen *et al.* (1997a). These results were kindly provided in electronic form for regraphing.

occurs when the atmospheric loading of natural aerosols increases, even though the sign of the forcing in this case is different between low latitudes and polar regions (Harvey, 1988). Furthermore, similar response patterns are seen for a variety of past climates, both warmer and colder than the present climate (Section 8.1 and Figure 8.4). Thus, for many cases, both the global mean temperature response and the regional temperature response are expected to vary in rough proportion to the global mean radiative forcing, and are largely independent of the geographical pattern of the forcing.

An exception to the above may occur if the total forcing is large enough that some critical threshold is crossed and the climate system abruptly jumps to some completely different state (having, for example, a completely different ocean circulation pattern). In this case, the regional climate patterns and probably even the global mean temperature response would no longer vary in proportion to the global mean forcing. Furthermore, when the regional variation in the forcing is sufficiently large (and different for different forcings), the regional response patterns will depend on the specific forcings involved and not just on the global mean forcing. This is the case for anthropogenic aerosols, which are highly concentrated spatially.

6.7 Radiative forcings in the 1990s

Table 6.2 summarizes estimates of the radiative forcings due to the changes in concentration of GHGs and aerosols between the pre-industrial period and the mid-1990s. The main source of uncertainty in the radiative forcing due to changes in stratospheric ozone arises from uncertainty in the vertical profile of changes within the stratosphere. In the case of tropospheric ozone, the main uncertainty results from uncertainty in the change that has occurred since pre-industrial times. The computation of the radiative forcing due to anthropogenic aerosols is undoubtedly the most difficult and uncertain of all the anthropogenic forcings. First, there is easily a factor of 2 uncertainty concerning just how much the concentration of key aerosols has increased since pre-industrial times. Second, there is even greater spatial variability in the changes in aerosol concentration than for tropospheric ozone. Third, and as previously noted (Section 2.5), the aerosol forcing involves both direct and indirect effects. The direct effect involves the reflection of solar radiation by the aerosol particles themselves, while the indirect effect involves modifications in the properties and/or lifespan of clouds. Fourth, the direct effect depends on the size distribution of the aerosol particles (which is somewhat uncertain), on the presence of impurities, and when impurities are present, on the geometry of the impurities within the main aerosol substance. The indirect effect is also complicated, depending as it does on microphysical processes in clouds and the presence of alternative sources of cloud condensation nuclei.

The total GHG radiative forcing in the 1990s is estimated to be $2.1-3.2\,\mathrm{W}$ m^{-2}, with most of the uncertainty due to uncertainty in the ozone forcing. This is equal to the forcing that would be caused by a CO_2 increase alone by about 50–80%. The direct sulphur + black carbon forcing is estimated to be 0.1 to $-0.4\,\mathrm{W\,m}^{-2}$ (a slight cooling effect), while the indirect sulphate forcing is estimated to be -0.5 to $-2.2\,\mathrm{W\,m}^{-2}$ if extreme estimates are excluded. Organic carbon and nitrate aerosols are, together, estimated to have caused a forcing of $-0.25\,\mathrm{W\,m}^{-2}$. The total global mean aerosol forcing is large enough, in principle, to have offset a large portion of the heating due to increases in GHGs. In regions subject to the largest aerosol loadings, however, the cooling perturbation could be 2–3 times the GHG heating.

Table 6.2 Summary of the radiative forcing due to changes in the composition of the atmosphere between the pre-industrial period and the mid-1990s. The direct radiative forcing due to methane is based on the observed increase, and so includes the effect of the change in the OH concentration induced by the anthropogenic emission of methane. The indirect methane forcing due to tropospheric O$_3$ is the contribution of the methane increase to the inferred increase in tropospheric O$_3$, so it is partly redundant with the tropospheric O$_3$ forcing that is shown further down in the table

Forcing	Magnitude (W m^{-2})	References
CO$_2$	1.40	Harvey (1999)
CH$_4$, direct	0.44	
tropospheric O$_3$	(0.11)	} Lelieveld *et al.* (1998)[a]
stratospheric H$_2$O	0.02	
N$_2$O	0.14	Harvey (1999)
CFCs	0.32	Harvey (1999)
Other well-mixed GHGs	0.05	Harvey (1999)
Total for well-mixed GHGs	2.37 ± 0.24[b]	
Increase of tropospheric O$_3$	0.3 to 0.7	Harvey (1999)
Loss of stratospheric O$_3$	−0.2 to −0.3[c]	Forster and Shine (1997), Hansen *et al.* (1997a)
	−0.1[d]	Forster *et al.* (1997)
Total GHGs	2.1 to 3.2	
Combined S and C aerosols, direct effect	0.1 to −0.4	Harvey (1999)
S aerosol, indirect effect	−0.5 to −2.2	Harvey (1999)
Organic carbon aerosols	−0.22	Hansen *et al.* (1998)
Anthropogenic dust	−0.12	Hansen *et al.* (1998)
Nitrate aerosols	−0.03	van Dorland *et al.* (1997)
Total aerosols	−0.8 to −3.0	

[a] As an indication of the uncertainty, note that Minschwaner *et al.* (1998) computed an adjusted direct forcing for 1750–1992 of 0.55 W m^{-2}.
[b] ±10% uncertainty has been arbitrarily added to the total forcing.
[c] Based on the radiative forcing at the conventionally defined tropopause, for the period 1979–1994. Very little if any ozone loss occurred prior to 1979.
[d] Obtained by scaling the above forcing by a factor of 0.33 in order to give the forcing at the radiative tropopause. See text for clarification.

6.8 Concept of an equivalent CO$_2$ increase

Because the global mean temperature response to radiative forcing perturbations involving a mixture of GHG increases depends only on the global mean forcing (to within ±20%), it is useful to determine the increase in CO$_2$ concentration alone that would give rise to the same global mean forcing as the particular combination of GHG increases that actually occurs. This is referred to as the *equivalent CO$_2$ increase*. It is not appropriate to include forcings for which the climate sensitivity is different from that for a CO$_2$ increase, nor is it recommended to include forcings with a strong enough regional variation to produce a markedly different response pattern than for increases in CO$_2$ or other well-mixed GHGs. Thus, the global mean radiative forcing due to increases in tropospheric aerosols and O$_3$ should not be lumped in with other radiative forcings before computing the equivalent CO$_2$ increase.

As an example, consider the radiative forcings shown in Table 6.2 for the increases in GHG concentrations that had occurred by 1995. The CO$_2$ concentration had increased by 28% and caused an adjusted radiative forcing of 1.40 W m^{-2}. However, the total radiative forcing from well-mixed gases by 1995 was 2.37 W m^{-2}, which is what would be produced by CO$_2$ alone if it were to increase in concentration by 55%. The equivalent CO$_2$ increase is thus 55%, and the equivalent CO$_2$ concentration in 1995 was 430 ppmv.

6.9 Summary

To summarize this chapter, the response of surface temperature to a perturbation in the radiative fluxes is governed by the change in net radiation at the tropopause, after allowing for adjustment of stratospheric temperatures. This change in net radiation is referred to as the adjusted radiative forcing. The climate sensitivity is the ratio of the global mean temperature response to the adjusted global mean radiative forcing. Although the climate sensitivity is not strictly constant, the change in its strength over a modest range of forcing magnitudes is likely to be quite small compared to the underlying uncertainty (a factor of 3) in the magnitude of the climate sensitivity. The global mean temperature response and its spatial structure do not depend to any great extent on the spatial structure of the heating perturbation for changes in the concentration of well-mixed gases, in solar luminosity, or in natural aerosols. Rather, the global mean temperature response and its spatial variation are governed by the global mean radiative forcing and the feedbacks that come into play. For forcings involving large increases in non-absorbing aerosols, the geographical pattern of the temperature response (but not the global mean temperature response) depends on the geographical pattern of the forcing. For increases in absorbing aerosols and for changes in stratospheric and tropospheric O$_3$, the climate sensitivity appears to be different from those of other forcings, owing to the fact that different cloud feedbacks come into play. However, in this case the possible differences in the climate sensitivity between aerosol or ozone increases and other forcing factors are small compared to the enormous uncertainty in the aerosol or ozone forcing.

Questions for further thought

1. Explain why the Earth's atmosphere has a "greenhouse" effect, that is, why the presence of the atmosphere reduces the net emission of infrared radiation to space.

2. In computing climate sensitivity (the ratio of temperature response to radiative forcing), the induced change in the atmospheric water vapour content as the climate changes is accounted for, but possible induced changes in the concentrations of other greenhouse gases are not accounted for. Why is this distinction made?

3. Compare and contrast the concepts of radiative forcing and radiative damping. How is the temperature response related to each?

4. Rank or classify the various radiative forcings in the 1990s according to (a) magnitude, (b) sign, and (c) degree of uncertainty.

5. What is the practical application of the assumption that the climate sensitivity is constant? Under what conditions is the assumption valid or invalid, and why?

Chapter 7

Response of the carbon cycle and other biogeochemical cycles: translating emissions of GHGs and aerosols into concentrations and radiative forcing

During the 1980s an average of 5.5 ± 0.5 Gt (gigatonnes, or billions of tonnes) of carbon were released to the atmosphere every year as CO_2 due to the burning of fossil fuels, with an additional 1.6 ± 1.0 Gt C per year due to tropical deforestation and other land use changes (Schimel *et al.*, 1996). Regrowth of forests on abandoned agricultural land in the NH mid-latitudes might have absorbed 0.5 ± 0.5 Gt C per year, to give a range in total net anthropogenic emissions of 4.6–8.6 Gt C per year. Atmospheric CO_2 increased at the rate of 3.3 ± 0.2 Gt C per year during this time, which implies that 1.1–5.5 Gt C per year, or about one-third to two-thirds of total anthropogenic emissions, are quickly removed from the atmosphere at present. This carbon is thought to be removed primarily through absorption by the oceans and by an increase in the rate of global photosynthesis due to the stimulatory effect of the 25% increase in atmospheric CO_2 which has occurred since the Industrial Revolution. Recent analysis of oceanic and atmospheric data indicate that the likely rate of oceanic uptake is 2.0 ± 1.0 Gt C per year (Section 7.8). Thus, a consideration of the overall CO_2 budget implies that the terrestrial biosphere is absorbing an extra 0–4.5 Gt C per year.

Analysis of the changing ratios of different carbon isotopes in the atmosphere reduces this uncertainty somewhat. These ratios are affected by combustion of fossil fuels and a reduction in the mass of the terrestrial biosphere, and indicate that there has been little change in the total mass of the biosphere in recent decades (Siegenthaler and Oeschger, 1987). This implies that the undisturbed biosphere has been absorbing an amount of carbon comparable to the net release as a result of deforestation + regrowth on abandoned land. This could be a rate of absorption as large as 2.6 Gt C per year. Inasmuch as the global net primary productivity (or NPP, equal to plant photosynthesis minus plant respiration) is on the order of 50–60 Gt C per year, the stimulation required to balance the carbon budget is no more than about 5%, which will be difficult to detect through direct measurement, particularly since part

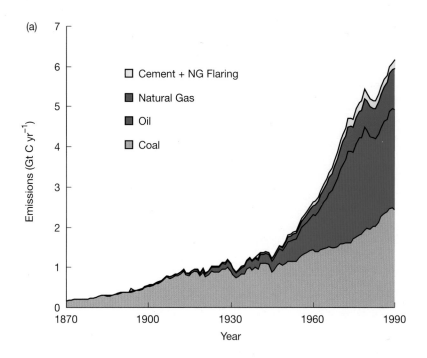

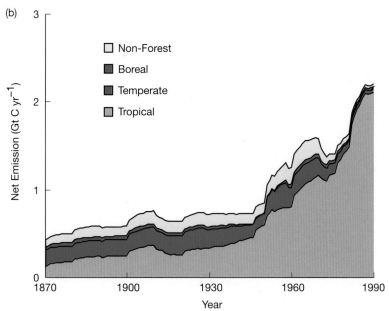

Plate 1 (a) Variations in CO_2 emissions (Gt C yr^{-1}) from 1870 to 1990 due to the combustion of coal, oil, natural gas, and the sum of the three. (b) Estimates net emissions of CO_2 (Gt C yr^{-1}) from 1870 to 1990 due to land use changes involving tropical, temperate, and boreal forest; grasslands; and the sum of the four. Fossil fuel and cement emission data are from Maryland *et al.* (1998) and were obtained from the web site http://cdiac.esd.ornl.gov, while land use emission estimates are from Houghton (1998) and were kindly provided in electronic form.

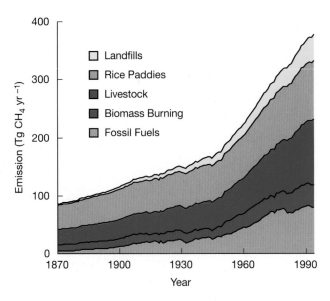

Plate 2 Historical variation in anthropogenic CH_4 emissions (Tg CH_4 yr^{-1}) from 1870 to 1990, as estimated by Stern and Kaufmann (1996a) and given at the web site http://cdiac.esd.ornl.gov/ftp/trends.ch4_emis/ch4.dat.

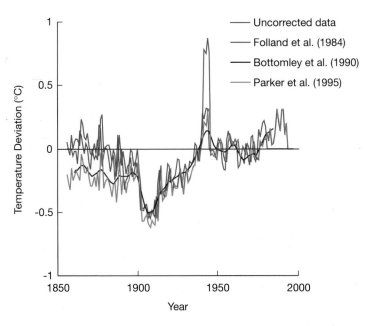

Plate 3 Uncorrected, global mean NMAT anomalies (°C) (green line) and NMAT anomalies as corrected by Folland *et al.* (1984) (blue line), Bottomley *et al.* (1990) (purple line), and Parker *et al.* (1995) (red line). Based on data files kindly provided by the UK Meteorological Office. The available Bottomley *et al.* (1990) data were smoothed.

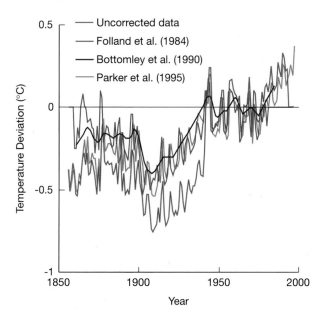

Plate 4 Variation in uncorrected global mean SST anomalies (°C) and as corrected by Folland *et al.* (1984), Bottomley *et al.* (1990), and Parker *et al.* (1995). Based on data files kindly provided by the UK Meteorological Office. The available Bottomley *et al.* data were smoothed.

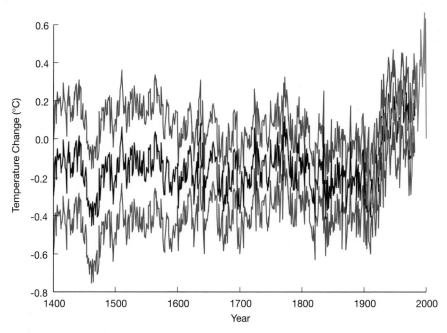

Plate 5 Variation in NH mean surface air temperature, as reconstruted by Mann *et al.* (1998) based on various proxy climate indicators. Also shown are the ±2 standard deviation error bars, (light blue lines) and observed NH mean temperature variation since 1990 (red line). Based on data obtained from the national Oceanographic and Atmospheric Administration (NOAA) Paleoclimatology web site http://www.ngdc.noaa.gov/paleo/pubs/mann1998.

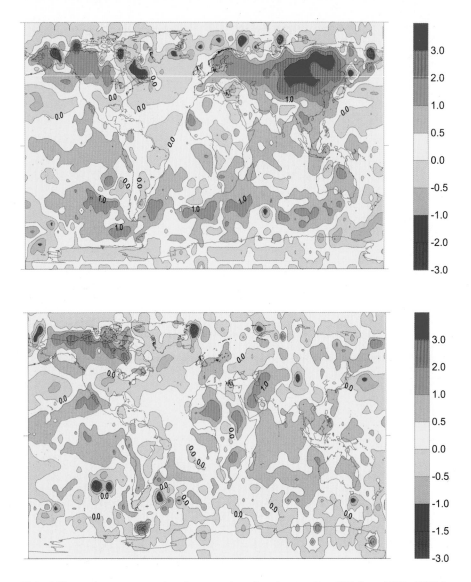

Plate 6 Change in mean annual and seasonal surface temperature (°C) from 1950–1959 to 1990–1998. Based on data obtained from the Climate Research Unit of the University of East Anglia (web site http://www.cru.uea.ac.uk).

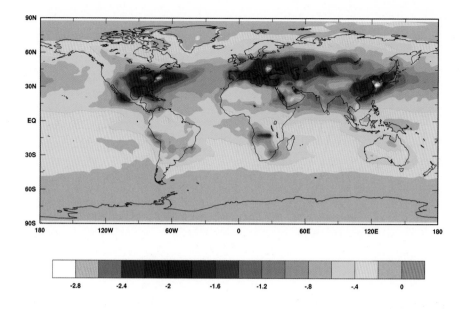

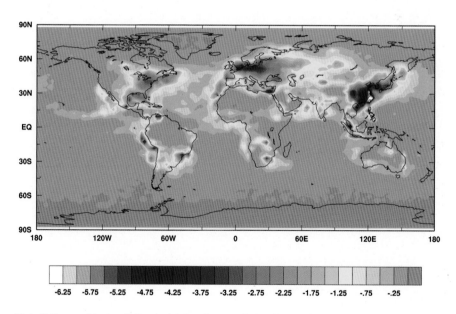

Plate 7 Geographical variation in (a) the direct radiative forcing due to sulphate aerosols, and (b) the indirect radiative forcing due to sulphate aerosols, as simulated by Kiehl *et al.* (1999). Reproduced from Kiehl *et al.* (1999).

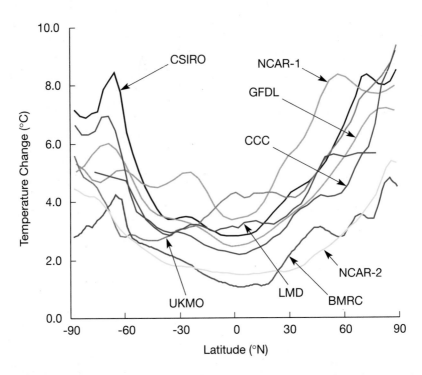

Plate 8 Variation with latitude in the zonally average steady-state surface-air temperature change for a CO_2 doubling, as simulated by various AGCMs. All results are based on data files directly provided to the author by the research institutes shown on this figure, but correspond to the following publications: CCC, Boer *et al.* (1992); CSIRO, Watterson *et al.* (1997), their Mark-2 version; GFDL, Manabe *et al.* (1991); LMD, Le Treut *et al.* (1994); NCAR-1, Washington and Meehl (1993); NCAR-2, Jeff Kiehl (personal communication, 1998); UKMO, Johns *et al.* (1997).

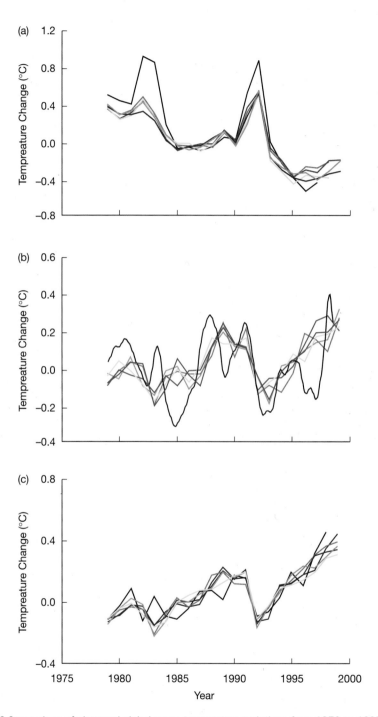

Plate 9 Comparison of observed global mean temperature variations from 1970 to 1995 and as simulated by Hansen *et al.* (1997c) for (a) the stratosphere, (b) the troposphere, and (c) the surface. For the troposphere, the MSU-2D data as corrected by Wentz and Schabel (1998) are shown.

of the increased photosynthesis is likely to go into an increased mass of roots and soil carbon.

The biosphere and oceans, because they act to remove anthropogenic CO_2, are referred to as carbon "sinks". In order to project atmospheric CO_2 concentrations associated with future CO_2 emissions, it is important to know the relative importance of the terrestrial biosphere and oceanic sinks today, and how each sink will respond in the future. These sinks can be expected to change both as a result of continuing injections of CO_2 from human activities into the atmosphere, and as a result of the climatic changes – temperature and precipitation in particular – induced by the very buildup of CO_2 and other greenhouse gases. The terrestrial biosphere sink is not infinite, and there is widespread agreement that it is of a temporary nature because terrestrial ecosystems (including soils) do not have an unlimited capacity to absorb carbon. Furthermore, the uptake of CO_2 by the terrestrial biosphere is not irreversible. On the other hand, the oceanic sink can be regarded as almost infinite (relative to the size of the fossil fuel resource) and close to irreversible, although it too could weaken considerably. The CO_2 that is not removed by these (or any other) sinks will accumulate in the atmosphere.

In this chapter we discuss the many ways in which the carbon cycle might respond to further increases in atmospheric CO_2 concentration and to concurrent climatic change. We then discuss the techniques used to estimate what the current rates of uptake by the terrestrial biosphere and ocean are, and review the resultant estimates. In the case of the terrestrial biosphere sink, there is the additional issue of attribution – the estimated sink strength during recent decades could be due to the direct effect of higher atmospheric CO_2 in stimulating photosynthesis, but it could also be due, at least in part, to the effect of fertilization by nitrogen pollution or to the effect of short-term climatic variability.

7.1 Response of the terrestrial biosphere to higher atmospheric CO_2 in the absence of climatic change

The important processes involving the terrestrial component of the carbon cycle are photosynthesis, respiration by plants themselves for growth (growth respiration) and to maintain the living plant processes (maintenance respiration), and respiration by micro-organisms that decompose plant litter (heterotrophic respiration). The decomposition of plant litter releases nutrients that can be used to support new plant growth. Plant litter that is particularly resistant to decomposition can accumulate in the soil, locking up nutrients but also storing carbon on a long-term basis. Plant photosynthesis is usually stimulated by higher atmospheric CO_2 concentration – a phenomenon sometimes referred to as *CO$_2$ fertilization*. Higher CO_2 concentration can also enhance or inhibit rates of maintenance respiration. An increase in photosynthesis implies an increased transfer of carbon from the atmosphere to the biosphere, thereby tending to decrease atmospheric CO_2 concentration and increasing carbon storage on land. An increase in respiration has the opposite effect. Carbon dioxide enters the leaves of plants for use in photosynthesis through small openings called *stomata*. An increase in the CO_2 concentration allows smaller stomatal openings, thereby reducing the loss of water through evapotranspiration. This can indirectly affect the carbon balance by extending the active growing season where water is a limiting factor, and by changing the rate of decomposition of soil organic matter by reducing the drying of soils in summer.

Initial, direct effects on the rate of photosynthesis

Photosynthesis in a given species of plant occurs through one of three different chemical pathways. In one pathway, an intermediate compound with three carbon atoms occurs, so this pathway is referred to as the C_3 pathway, and plants using it are called C_3 plants. In the second pathway, referred to as the C_4 pathway, an intermediate compound with four carbon atoms occurs. The third pathway is not common and so is not of concern to us here. All trees and most temperate- and high-latitude grasses are C_3, while tropical grasses are C_4. Photosynthesis is a chemical reaction that is powered by sunlight and catalysed by the ribulose biphosphate carboxylase (rubisco) enzyme when a CO_2 molecule attaches to it. However, if an O_2 molecule attaches to the rubisco enzyme, then respiration can occur in the presence of sunlight. This respiration – known as *photorespiration* – undoes part of the effort of photosynthesis, thereby reducing the net productivity of the plant. The ratio of photosynthesis to photorespiration thus depends on the ratio of CO_2 to O_2 molecules bombarding the rubisco enzyme (and on the reaction probabilities). In C_4 plants, which evolved around or prior to 8 million years ago (Ehleringer *et al.*, 1991) as the atmospheric CO_2 concentration fell, the CO_2 concentration around the rubisco enzymes is kept at a higher level through an internal pumping mechanism, thereby inhibiting photorespiration.

An increase in the concentration of atmospheric CO_2 will increase the probability of a CO_2 molecule binding with the rubisco enzyme in C_3 plants, thereby reducing the rate of photorespiration and increasing net photosynthesis. This effect is negligible in C_4 plants because they already have a high CO_2 concentration around the rubisco enzymes. Hundreds of experiments with dozens of C_3 plant species indicate that plant growth increases by 30–40% (and sometimes much more) when the atmospheric CO_2 concentration is doubled (Idso and Idso, 1994).

Most of the experiments measuring the effect of higher atmospheric CO_2 on plants were performed in controlled and somewhat artificial greenhouse settings, although in recent years an increasing number of experiments have been performed in which the CO_2 concentration in outside air is increased (by piping CO_2 through the site and releasing it at a controlled rate) and the response of plants under closer-to-natural conditions observed. These latter experiments are referred to as "free-air CO_2 enhancement" experiments or FACE experiments. However, only a few studies with elevated CO_2 have used wild plants growing in natural ecosystems, and only a handful have lasted more than two growing seasons.

Role of environmental variables

Reviews by Gifford (1992) and Idso and Idso (1994) indicate the following relationships for the short-term (a few weeks to months) response of plants to higher CO_2:

- the percentage enhancement in growth is greater under water-limited conditions than when water is not in short supply;
- the percentage enhancement in growth is greater the warmer the temperature, but below 12°C for carrot and radish, an increase in CO_2 concentration reduces plant growth;
- the percentage enhancement in growth can be higher, smaller, or unaltered under conditions of low nutrient availability compared to conditions of high nutrient availability;
- the percentage enhancement in growth can be smaller or larger under conditions of low light compared to high light; and
- the percentage enhancement in growth under polluted conditions is greater than in non-polluted conditions.

Although the percentage enhancement of growth is often greater under stressed conditions (low light, low nutrients, low water, or high pollution) than under unstressed conditions, the absolute increase in the rate of growth is usually smaller because the initial growth rate in ambient CO_2 is smaller. Nevertheless, photosynthesis is able to increase under higher CO_2 even when other factors (such as light, water, and nutrients) are limiting, because higher CO_2 lets the plant make more efficient use of limited resources.

Law of diminishing returns

The short-term response of plant photosynthesis shows a tendency to saturate; that is, each successive increase in atmospheric CO_2 has a progressively smaller stimulatory effect on the rate of photosynthesis. This is illustrated in Figure 7.1, which shows the average enhancement in short-term plant growth as CO_2 concentration increases, for resource-limited or stressed conditions, and for non-resource-limited and stress-free conditions.

Biochemical downregulation of the photosynthetic response

Many, but not all, of the plant species that show a marked increase in the rate of photosynthesis upon initial exposure to higher CO_2 show a gradual reduction (by one-third to two-thirds) in the stimulatory effect of higher CO_2 over time (Bazzaz, 1990). This reduction in the initial stimulation of photosynthesis is referred to as *downregulation*. Downregulation occurs for two known reasons. First, the plant might not be able to use the additional products of photosynthesis because plant growth is still strongly limited by environmental constraints. Second, the concentration of the rubisco enzyme – which catalyses the photosynthetic reaction – decreases under higher CO_2. Since about 25%

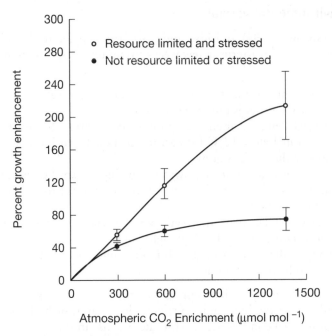

Figure 7.1 Average percentage enhancement in plant growth as the magnitude of the enhancement in atmospheric CO_2 concentration changes. Results are shown for two sets of conditions: (i) for plants with less than adequate light, water, and nutrients, or experiencing stress induced by high levels of salinity, air pollution, or temperature (o); and (ii) for plants experiencing none of the above (●). Reproduced from Idso and Idso (1994).

of plant nitrogen is found in the rubisco enzyme, this means that a smaller amount of nitrogen is required for a given rate of photosynthesis. This in turn implies that the *nutrient utilization efficiency* (NUE) increases as downregulation occurs. As a result, ecosystem-scale productivity might not be as adversely affected by downregulation as is the productivity of an individual leaf or plant, particularly if a greater density of plant growth occurs.

Interactions with soil microfauna and soil nutrients

A number of feedbacks involving nutrient cycling could significantly alter the direct or initial effects of higher CO_2 (Zak *et al.*, 1993; Diaz *et al.*, 1993; Gifford, 1992, 1994). Some of these could reduce the stimulatory effect of higher CO_2 on plant growth and carbon storage, while others could enhance it. The complexity of plant–soil–nutrient interactions is so great, and there are so many opposing negative and positive feedbacks, that the net effect of these feedbacks on carbon storage – and hence on the buildup of atmospheric CO_2 – is likely to vary from ecosystem to ecosystem and possibly even within a given ecosystem. The net feedback might also change over time.

Other considerations involving the photosynthetic response

Changes in the partitioning of plant production among different plant components as overall productivity increases can also strongly modulate the net increase in plant biomass and are likely to reduce the increase in storage (Post et al., 1992; Norby et al., 1992; Luo et al., 1996). Changes in the rate of senescence (leading to leaf fall) can also reduce the end-of-season carbon gain (McConnaughay et al., 1996). A further important consideration in the net ecosystem response to higher atmospheric CO_2 is the effect of changes in the relative abundance of different plant species due to the differential effects of higher CO_2 on different species. Allowance for changes in species abundances could enhance the increase in carbon storage due to higher CO_2 by as much as 30% (Bolker et al., 1995) as long as co-occurring changes in climate do not require shifts in species distributions. When climatic changes are large enough to require a shift in species distributions, a temporary (centuries) reduction in carbon storage could occur due to the lag between decline of maladapted species and the arrival of new species at a given site (Section 7.2).

Decreased stomatal conductance

During photosynthesis, the stomata need to be open to permit more CO_2 to enter the leaf; this opening is associated with a loss of water from leaves. With a higher atmospheric CO_2 concentration, a smaller stomatal opening is required for a given inflow of CO_2, which in turn reduces the water loss. In a survey of experiments on 23 tree species, Field et al. (1995) found an average decrease in stomatal conductance of 23% in response to a doubling of atmospheric CO_2. The water use efficiency (WUE), or ratio of carbon uptake to water loss, therefore increases. In cases where the availability of water limits growth, this can increase the amount of photosynthesis over the course of a growing season. This effect is applicable to both C_3 and C_4 plants.

Changes in the rate of respiration

A doubling of atmospheric CO_2 concentration reduces the rate of growth and maintenance respiration per unit of tissue dry mass by about 20% on average (Drake et al., 1996), although opposite effects have also been observed (Ryan, 1991; Wullschleger et al., 1994). A suppression of respiration has been reported for leaves, roots, stems, and even soil bacteria. There is still considerable debate as to whether there is a true direct suppression of respiration by elevated CO_2 concentration.

7.2 Effect of climatic change on the terrestrial biosphere sink ___

In the preceding section we discussed some of the myriad of ways in which an increase in atmospheric CO_2 can directly alter the net flow of carbon into or out of terrestrial ecosystems. Changes in climate, induced by the increase in CO_2

and other GHGs, will also alter the net uptake of CO_2 by the terrestrial biosphere, either by changing the rates of photosynthesis and respiration in healthy ecosystems, or by stressing existing ecosystems to the point that they collapse and are replaced by new ecosystems. During the transition from one forest ecosystem to another, forest dieback and an increase in the frequency of forest fires might occur, which would inject additional CO_2 into the atmosphere.

Equilibrium effects

Higher temperatures tend to increase plant maintenance respiration and the respiration of litter and soil organic matter. Large amounts of carbon are stored in the soils at high latitudes, which are also the regions that are expected to experience the greatest warming (Section 8.2). This raises the prospect of significant releases of carbon from these regions (Oechel *et al.*, 1993). Two factors could greatly reduce the loss of plant and soil carbon due to warmer temperatures. First, an increase in the respiration of soil organic matter would increase the availability of nutrients. Soil respiration is the process by which soil microorganisms digest plant carbon and burn it for their own energy needs, releasing certain nutrient-containing compounds in the process. This increases the availability of nutrients to the plant roots, so any increase in the rate of soil respiration would tend to increase photosynthesis rates under nutrient-limited conditions (independently of any increase due to higher CO_2 and warmer temperatures). This would further reduce the loss of soil organic matter, since there would be a larger flow of litter into the soil. Second, warmer temperatures could suppress the downregulation of the photosynthetic response to higher CO_2 that would otherwise occur. As noted in the previous section, the initial stimulation of photosynthesis seen when CO_2 is doubled tends to decrease after a few years, and can completely disappear. This is probably due in part to the inability of plants to use the increased products of photosynthesis. However, warmer temperatures increase the demand for photosynthetic products, and are therefore expected to reduce the long-term downregulation of photosynthesis. As a result of these two considerations, it is hard to see how warming alone could cause a major pulse of carbon from the terrestrial biosphere to the atmosphere. However, an increase in the frequency of drought in the interiors of some continents is expected as the climate warms, for reasons that are well understood (Section 8.3). This could lead to large-scale dieback of the current vegetation, and, as discussed next, is the major reason for concern regarding a temporary carbon pulse as ecosystems are reorganized.

Transient effects

A potentially significant terrestrial biosphere–climate feedback could occur if and when climatic change becomes large enough to induce changes in the species composition at a given site. The lag between dieback of maladapted ecosystems and their replacement by new ecosystems as climate changes could lead to significant decadal-to-century time-scale decreases in terrestrial carbon storage, even in cases where the net effect of CO_2 and climatic changes would be to increase carbon storage once the species distribution is fully adjusted to the new climate (Solomon, 1986; Pastor and Post, 1988). Calculations by Dixon *et al.* (1994) indicate that the average flux over the next century due to climate-induced forest dieback could be as high as 4.2 Gt C yr^{-1}, and would be further increased to as high as 6.0 Gt C yr^{-1} if climate-induced expansion of agriculture into forested areas is included. However, potential CO_2-fertilization effects are excluded from these calculations. Furthermore, an increase in atmospheric CO_2 increases the optimal temperature for plant growth (Idso and Idso, 1994), so that the need for ecosystems to shift poleward as the climate warms will be reduced. Nevertheless, it is clear that, with unrestrained GHG emissions and global warming, there is a growing risk that what is probably a net carbon sink at present could significantly weaken and possibly become a net carbon source.

7.3 Estimation of the current rate of uptake of anthropogenic CO_2 by the terrestrial biosphere

As noted in the introduction to this chapter, a consideration of the known anthropogenic emissions of CO_2, the observed rate of atmospheric increase, and the estimated rate of uptake of CO_2 by the oceans, implies that the terrestrial biosphere (excluding the effects of land use changes) needs to be absorbing up to 4.5 Gt C per year in order to balance the global carbon budget. This upper limit of 4.5 Gt C per year is comparable to what would be expected based on a simple-minded extrapolation of the short-term response of plants to higher CO_2 as measured under controlled conditions in greenhouses or in the field. These results imply an increase in photosynthesis of 30–40% for a doubling of CO_2, and hence an increase of 8–10% – or 5–6 Gt C yr^{-1} – due to the 25% increase in the atmospheric concentration of CO_2 that has occurred so far. A more refined estimate is provided by Gifford (1994), who used a globally aggregated biosphere model driven by global mean changes in temperature and CO_2 to determine the current rate of uptake. Assuming a long-term increase in carbon storage of 10–40% for a CO_2 doubling and plausible lower and upper limits for the effects of temperature on photosynthesis and respiration, he obtained a net biomass + soil carbon uptake today of 0.5–4.0 Gt C yr^{-1}. These results largely span the range needed to balance the carbon budget. However, various additional observations, to be discussed next, can be used to show that the current rate of absorption by the terrestrial biosphere is substantially less than the upper limit of 4.5 Gt C per year that is obtained from a consideration of the global-scale budget. This in turn provides the clearest evidence that the 30–40% stimulation of photosynthesis found in short-term experiments cannot be simply extrapolated to the entire natural biosphere.

There are small spatial (and seasonal) variations in the concentration of CO_2 in the atmosphere that can be used to infer the geographical distribution and magnitude of the various sources and sinks of CO_2. For example, the current CO_2 concentration is about 3 ppmv greater north of 30°N than it is south of 30°S. Carbon occurs as two stable isotopes (varieties): 99% as ^{12}C and about 1% as ^{13}C (carbon also occurs as a radioactive isotope, ^{14}C, but ^{14}C accounts for only one carbon atom in about 10^{12}). Photosynthesis preferentially uses ^{12}C over ^{13}C, thereby enriching the remaining atmospheric CO_2 in ^{13}C, while the burning of fossil fuels and biomass releases ^{13}C-depleted CO_2 into the atmosphere. The absorption of CO_2 by the oceans, in contrast, entails relatively little discrimination between different isotopes. Information on the spatial distribution of the isotope ratio in the atmosphere can be used, alone or in conjunction with CO_2 concentration data, to infer the global magnitude of the terrestrial biosphere sink and its crude geographical variation. Such calculations are referred to as *inverse* calculations, because one starts with observations of the spatial variation in atmospheric concentrations and works backward to the required distribution of emissions.

Using data on the spatial variation in the carbon isotope ratio, Keeling *et al.* (1995) and Francey *et al.* (1995) estimated the biospheric exchange – the net CO_2 source resulting from human-induced land use emissions minus uptake due to CO_2 fertilization and other causes. Keeling *et al.* (1995) deduced an average exchange from 1978 to 1994 of 0.21 Gt C yr^{-1}, while Francey *et al.* (1995) deduced an average exchange from 1982 to 1992 of about 1.5 Gt C yr^{-1}. The discrepancy between the two results is largely due to the use of isotope data from different monitoring stations in the two studies. Both groups deduced significant interannual variations in the net exchanges, but disagree significantly during the period of overlap.

Global mean atmospheric CO_2 and $\delta^{13}C$ variations have also been used to infer the variation in the net exchange of carbon between the terrestrial biosphere and the atmosphere during the past 200 years or so. Figure 7.2 shows the net exchange since 1850 as reconstructed by Siegenthaler and Oeschger (1987). The atmospheric $\delta^{13}C$ variations were inferred from measurements of the isotope ratio of tree rings. Of particular interest is the inference that the net exchange has been very close to zero since about 1980, which supports the exchange estimate of Keeling *et al.* (1995), cited above. Also shown in Figure 7.2 is the global net emission due to changes in land use, as calculated by Houghton (1998) and previously shown in Plate 1(b). The difference between the two gives the required terrestrial biosphere sink.

Different analyses of the geographical variation in atmosphere ^{13}C lead to the conclusion that there is a large sink in NH mid-latitudes. Inverse calculations by Tans *et al.* (1990), Ciais *et al.* (1995), and Fan *et al.* (1998) place most of the mid-latitude sink on land rather than in the oceans. Indeed, Fan *et al.* (1998) calculate that most of the NH mid-latitude sink is due to uptake by forests in North America (1.7 ± 0.5 Gt C yr^{-1}), with Eurasia–North Africa contributing only 0.1 ± 0.6 Gt C yr^{-1}. The CO_2 source deduced by Ciais *et al.* (1995) for tropical regions (16°S–16°N) is smaller (1.5 Gt C yr^{-1}) than expected from oceanic outgassing and tropical deforestation, which implies some absorption of CO_2 by the undisturbed tropical biosphere. This is supported by Fan *et al.* (1998), who deduced a net tropical land source of only 0.2 ± 0.9 Gt C yr^{-1}.

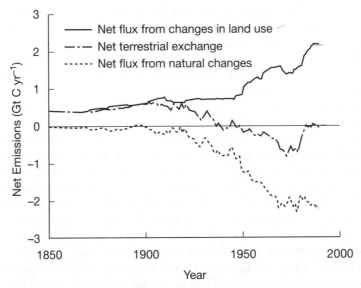

Figure 7.2 Net exchange of carbon between the atmosphere and the terrestrial biosphere as computed by Siegenthaler and Oeschger (1987), the net flux to the atmosphere due to land use changes as computed by Houghton (1998) and previously shown in Plate 1(b), and the difference between the two. Based on data kindly provided by R. A. Houghton.

The regional terrestrial sources and sinks inferred from these inverse calculations include the direct effects of human-induced land use changes (regrowth of forests on abandoned agricultural land in the NH mid-latitudes; deforestation and concurrent regrowth of forests on rapidly abandoned land in the tropics). The difference between the regional net source as deduced from inverse calculations, and the regional net source due to land use changes as estimated from land use and other ground-based data, would presumably reflect the regional CO_2 fertilization effect. However, as explained below, the CO_2 fertilization effect is only one of three factors that can explain the difference between land use-related fluxes and the net biospheric exchange of carbon.

7.4 Causes of the recent absorption of CO_2 by the terrestrial biosphere

Two factors other than CO_2 fertilization and regrowth of forests on abandoned land could be causing the terrestrial biosphere to act as a sink at present: fertilization of terrestrial ecosystems (primarily forests) with nitrogen from NO_x emissions (which also contribute to acid rain), and the effects of recent decadal-scale variability in climate.

With regard to the former, Melillo *et al.* (1996) estimate that fertilization by deposition of NO_x caused a global carbon sink of 0.5–1.0 Gt C yr^{-1}. This is comparable to lower estimates of the magnitude of the terrestrial biosphere sink needed to balance the global carbon cycle. There is evidence that decadal fluctuations in climate in the recent past have also affected the net storage of carbon in the terrestrial biosphere. According to calculations by Dai and Fung (1993), increases in NPP in middle and high latitudes due to warming, along with other changes elsewhere, created a carbon sink of 0.59 ± 0.15 Gt C yr^{-1}, averaged over the period 1940–1984. This calculated sink does not include possible effects of CO_2 and N fertilization.

As noted in the preceding section, calculations based on the atmospheric distribution of ^{13}C imply that there is a net terrestrial biosphere sink in NH mid-latitudes. This sink would include the effect of CO_2 and N fertilization, climatic change, and regrowth of forests on previously deforested land. Subtracting the carbon uptake due to regrowth of forests would then give the net effect of the first three factors. However, the carbon sink due to forest regrowth itself would include the effect of CO_2 and N fertilization, and of climatic change. Houghton (1996) presented evidence that part of the mid-latitude NH terrestrial biosphere sink is indeed due to CO_2 and/or N fertilization of regrowing forests, and not just due to normal regrowth following abandonment of previously cleared land. He compared the expected sink based on changes in land use, and based on direct measurements of the rate of growth of new forests. The former method gives the expected sink in the absence of fertilization effects, and amounts to 0.0 ± 0.5 Gt C yr^{-1}. The second method includes any effects of fertilization (and of climatic change), and yields an estimated sink of 0.8 ± 0.4 Gt C yr^{-1}. The implied fertilization–climate effect, 0.8 ± 0.9 Gt C yr^{-1}, is about half of the total terrestrial biosphere sink that is required to balance the global carbon budget. The net sink of 0.8 ± 0.4 Gt C yr^{-1}), is also only half of that inferred by Fan $et\ al.$ (1998) from inversion of atmospheric data (1.7 ± 0.5 Gt C yr^{-1}), but both estimates are probably more uncertain than is implied by the quoted uncertainty range.

The remainder of the carbon sink could be in the tropics and due to CO_2 fertilization effects. Recall, from Section 7.3, that a sink is expected in the tropics owing to the fact that the source deduced by inverse calculations (0.2 ± 0.9 Gt C yr^{-1}) is much less than would be expected based on estimated emissions due to changes in land use (1.6 ± 1.0 Gt C yr^{-1}). However, changes in climate have not caused a significant sink in the tropics according to Dai and Fung (1993), and NO_x deposition is also not likely to be important to the tropics. This leaves a tropical CO_2 fertilization effect of 0.0–3.0 Gt C yr^{-1}. Thus, CO_2 fertilization effects in the tropics could easily equal or exceed the combined CO_2–N fertilization effect in mid-latitudes.

7.5 The oceanic CO_2 sink in the absence of climatic change

The major processes involved in the oceanic part of the carbon cycle were identified in Section 2.4 and illustrated in Figure 2.7. These processes are gaseous exchange of CO_2 between the atmosphere and ocean; chemical equilibration between CO_2, CO_3^{2-}, and HCO_3^-; vertical transfers between the mixed layer and deeper ocean through the biological pump, advective overturning, convection, and diffusion; and the net burial of organic tissue and CaCO$_3$ in oceanic sediments. Owing to the action of the biological pump, the concentration of total dissolved inorganic carbon (DIC, the sum of [CO$_2$], [CO$_3^{2-}$], and [HCO$_3^{-}$]) in surface waters is about 10% less than in deep water, and since the atmosphere tends to equilibrate with the surface water, this causes the atmospheric CO$_2$ concentration to be much smaller than it would be in the absence of the biological pump.

In this section, the steps involved in the oceanic response to anthropogenic emissions of CO$_2$ are outlined, followed by a discussion of the role of climate–ocean carbon cycle feedbacks, and estimates of the current rate of absorption of anthropogenic CO$_2$ by the oceans.

Air–sea exchange of CO$_2$

Anthropogenic emissions of CO$_2$ lead to an increase in the atmospheric partial pressure of CO$_2$ (pCO$_2$), causing it to be greater than the average value in ocean surface water. As a result, there is a net flow of CO$_2$ from the atmosphere to the ocean. As CO$_2$ enters the mixed layer, the mixed layer CO$_2$ increases and the atmospheric CO$_2$ decreases. With no further emission of CO$_2$ into the atmosphere, the two would rapidly come into balance (within a year) and there would be no further flow of CO$_2$ into the ocean. The mixed layer pCO$_2$ depends on the CO$_2$ concentration, not on the total DIC concentration. Recall that 90% of DIC is as HCO$_3^{-}$, about 10% as CO$_3^{2-}$, and less than 1% as CO$_2$. A chemical equilibrium is established among these three components, through the net reaction

$$H_2O + CO_2 + CO_3^{2-} \rightleftharpoons 2HCO_3^{-} \tag{7.1}$$

which proceeds in both the forward and backward directions. The chemistry of sea water is such that, when CO$_2$ enters the mixed layer, it remains disproportionately in the form of CO$_2$. Specifically, when total DIC increases by 1%, the CO$_2$ concentration increases by 10%. As a result, the mixed layer pCO$_2$ increases 10 times faster than the increase in DIC. The ratio of the percentage increase in pCO$_2$ to the percentage increase in DIC is called the *buffer factor* and is designated by the symbol β (currently, $\beta \approx 10$). Since the rapid increase in mixed layer pCO$_2$ is a back-pressure that prevents further CO$_2$ from entering the ocean, it can be seen that this strongly limits the ability of the mixed layer to absorb CO$_2$ from the atmosphere. As shown in Box 7.1, the fraction of emitted carbon that can be absorbed by the mixed layer alone is $1/(1 + \beta)$.

We now have a somewhat paradoxical result: the chemistry of sea water, where 99% of the dissolved inorganic carbon is not in the form of CO$_2$, largely contributes to the present atmospheric CO$_2$ concentration being so low. However, this same chemistry – through the buffer factor – makes it difficult to put additional CO$_2$ into the mixed layer.

Unlike the terrestrial biosphere, *a higher concentration of CO$_2$ has almost no stimulatory effect on marine photosynthesis*, so there will be no direct effect of an

Box 7.1 Uptake of anthropogenic CO_2 by an atmosphere-mixed layer system

Consider a system consisting of a thoroughly mixed atmospheric box and an oceanic mixed layer box, which are in an initial equilibrium state with equal CO_2 partial pressures. The atmosphere pCO_2 is directly related to the mass of carbon in the atmosphere as CO_2, M_a, while the mixed layer pCO_2 depends on the $[CO_2]$ in water next to the air–sea interface. The amount of carbon in the mixed layer, M_{ML}, will depend on its depth, but assume that $M_{ML} = \alpha M_a$. A carbon mass ΔM is injected into the atmosphere as CO_2. Let f_a be the fraction of the injected CO_2 that remains in the atmosphere when a new steady state is established, so $f_{ML} = 1 - f_a$ is the fraction that is taken up by the oceanic mixed layer. In the new steady state, the atmospheric and oceanic pCO_2s will have increased by the same amount, so that we can write

$$\frac{\Delta(pCO_2)_a}{(pCO_2)_a} = \frac{\Delta(pCO_2)_{ML}}{(pCO_2)_{ML}} \tag{7.1.1}$$

The fractional change in atmospheric pCO_2 is equal to the fractional change in atmospheric carbon, while the fractional change in mixed pCO_2 is equal to β times the fractional change in mixed layer carbon, where β is the buffer factor. Thus, we can write

$$\frac{f_a\Delta M}{M_a} = \beta \frac{(1 - f_a)\Delta M}{M_{ML}} \tag{7.1.2}$$

Using $M_{ML} = \alpha M_a$, we obtain

$$f_a = \frac{\beta}{\alpha + \beta} \tag{7.1.3}$$

For $\alpha = 1$ and $\beta = 9.7$, which corresponds closely to the real oceanic mixed layer, $f_a = 0.91$ and $f_{ML} = 0.09$. Thus, the ocean ends up with only one-tenth as much new carbon as the atmosphere, due to the fact that $(pCO_2)_{ML}$ increases 10 times faster than $(pCO_2)_a$ as its carbon content increases.

increasing mixed layer CO_2 concentration on the strength of the biological pump. The strength of the biological pump is limited by the availability of nutrients, which are supplied by upwelling of deep water. Changes in the biological pump could therefore occur through climate-induced changes in the rate of upwelling, but not through the direct effect of higher CO_2 concentration in the mixed layer.

Downward mixing into the deep ocean

As the concentration of DIC increases in the ocean mixed layer, the downward fluxes due to advective and convective overturning will increase, since in both cases the sinking water parcels will have a greater DIC concentration than before. In addition, the upward diffusion of DIC will decrease, since the DIC concentration will now increase less rapidly with increasing depth (see Figure 2.8). We can think of the change in diffusion as a downward diffusive flux from the mixed layer, in response to higher mixed-layer DIC, superimposed on the pre-existing upward flux. Thus, the excess carbon in the mixed layer gradually works its way into the deeper ocean. On a global basis the dominant transfer process is diffusion, although at high latitudes, convective mixing is dominant.

Diffusion is a result of random turbulent eddies that mix water parcels and, in so doing, transfer dissolved constituents (and heat) from regions of high concentration to regions of low concentration. Vertical diffusion is a slow process. The time scale required for a perturbation to penetrate to a significant extent to a depth D is given by D^2/k, where k is the vertical diffusion coefficient. Given an effective k value of about $1.0 \times 10^{-4}\,\text{m}^2\,\text{s}^{-1}$ for DIC (see Harvey and Huang, 1999), the time scale is on the order of 100 years to penetrate 500 m. In contrast, the mixed layer equilibrates with the atmosphere on a time scale of about one year. As DIC penetrates into the deeper ocean, the mixed layer is able to absorb more CO$_2$ from the atmosphere, but because the rate of penetration below the mixed layer is so slow, downward mixing serves as a bottleneck on the oceanic uptake of anthropogenic CO$_2$. Again, the exception is in regions of deep convection, where carbon added to the mixed layer is rapidly mixed through several hundred to thousands of metres of water.

Complete mixing of the CO$_2$ that would be released from combustion of a large part of the recoverable fossil fuel resource (3800 Gt or 10% of the total carbon storage in the ocean) through the entire volume of the oceans would dilute the DIC of the mixed layer enough to allow the oceans to eventually absorb about 75–80% of the carbon added to the atmosphere, as shown in Box 7.2. The time required for most of this absorption to occur is 1000 years. However, partial dissolution of CaCO$_3$ sediments would allow the ocean to absorb much of the remaining CO$_2$, as explained below.

Role of oceanic sediments

As discussed in Section 2.4, biological productivity in the mixed layer produces a steady rain of CaCO$_3$ (calcium carbonate) particles. Most of these are in the form of calcite. The solubility of ocean water with respect to calcite increases with pressure and hence with depth, and the top 1–4 km of the ocean (depending on location) is supersaturated with respect to calcite, while the deep ocean is unsaturated. Figure 7.3(a) shows the geographical variation in the depth of the boundary between saturated and unsaturated water (the saturation horizon), while Figure 7.3(b) shows the percentage of CaCO$_3$ in seafloor sediments.

Box 7.2 Oceanic uptake of anthropogenic CO_2 after complete mixing

To estimate what fraction of anthropogenic CO_2 can eventually be taken up by the oceans, after thorough downward mixing of the anthropogenic perturbation has had a chance to occur, we can use Equation (7.1.3) of Box 7.1. The amount of carbon in the oceans is about 64 times the pre-industrial carbon content of the atmosphere (38,000 Gt/592 Gt), so $\alpha = 64$. Assuming for the moment that β remains fixed at 9.7, we obtain $f_a = 0.13$, so the ocean takes up 87% of the injected carbon. Thus, if 3800 Gt C of CO_2 are released, 500 will remain in the atmosphere and 3300 will be taken up by the ocean. This would increase the average DIC concentration by 8.6%, which would increase β to about 14. Using an average β of 12, our revised estimate of the oceanic uptake is 84% rather than 87% of the injected carbon. However, this assumes that the added carbon is eventually distributed uniformly throughout the ocean, which is not the case. Using the 1-D model of Harvey and Huang (1999), we find that about 1000 Gt C remain in the atmosphere, and the actual oceanic uptake is only 74% of the injected carbon. This is the amount of carbon that would be removed after 1000–2000 years; if 3800 Gt C were emitted during the next 100–200 years, the buildup of CO_2 during this period would be several times greater than the final accumulation of 1000 Gt C because of the bottleneck created by slow downward mixing of the anthropogenic perturbation below the mixed layer.

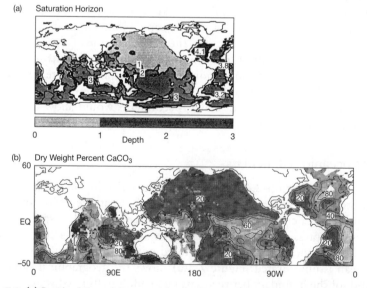

Figure 7.3 (a) Depth of the calcite saturation horizon in the world oceans, and (b) percentage of $CaCO_3$ in ocean sediments at the seafloor. Reproduced from Archer (1996a) and Archer (1996b), respectively.

As ocean water absorbs CO_2, $[CO_3^{2-}]$ decreases though reaction (2.4). This decreases the degree of supersaturation of water in the upper ocean with respect to $CaCO_3$. As CO_3^{2-}-depleted water penetrates into the deep ocean, the boundary between supersaturated and unsaturated water will rise, allowing the deepest $CaCO_3$ sediments to begin dissolving. As $CaCO_3$ dissolves, CO_3^{2-} is restored to the surrounding water. As this water mixes back up to the surface, the ocean can absorb more CO_2 from the atmosphere. The amount of $CaCO_3$ sediment that can dissolve is limited by the fact that, as dissolution occurs, the non-carbonate sediment is left behind, and this sediment eventually forms a capping layer thick enough to prevent further dissolution of carbonate sediments. Nevertheless, it is expected that enough carbonate sediment can dissolve to allow the oceans to take up about half of the remaining 20–25% of anthropogenic CO_2, so that the oceans ultimately take up 85–90% of CO_2 emitted into the atmosphere. However, several thousand years will be required for this to happen. Enhanced chemical weathering on land would take up the remaining 10–15%, but over a time period of about 200,000 years (Archer *et al.*, 1997).

7.6 Effect of climatic change on the oceanic sink

There are two main ways in which climatic change accompanying an increase in atmospheric CO_2 concentration could alter the oceanic carbon sink: through changes in ocean surface temperature, and through changes in the operation of the biological pump.

Ocean temperature feedback

Temperature directly affects the pCO_2 of sea water in two ways. First, the partitioning of total DIC between dissolved CO_2, CO_3^{2-}, and HCO_3^- depends on temperature. Second, and more importantly, the solubility decreases with increasing temperature. The net result is that the ocean is able to hold less CO_2 as the mixed layer warms, and there will be a counterflow of CO_2 back into the atmosphere.[1]

As a quantitative example, consider a cumulative emission of 4000 Gt C, which would be sufficient to double the pre-industrial CO_2 concentration after a period of about 1000 years (having fallen from much higher interim levels). If the global mean ocean surface temperature warming for a CO_2 doubling is 2°C, the atmospheric carbon content will be 1267 Gt rather than 1184 Gt (a difference of 7%), but the change in carbon content compared to pre-industrial times will be 675 Gt rather than 592 Gt – a difference of 14%. If the temperature change (prior to any pCO_2 feedback) is 4°C instead of 2°C, the CO_2 buildup is enhanced by 28% rather than 14%. In either case, the extra

---

[1] The oceanic warming below the mixed layer is irrelevant to this, since only the mixed layer is in direct contact with the atmosphere, and the net flow to or from the atmosphere depends on what happens to pCO_2 in this layer, not below.

CO_2 buildup would induce further warming that would induce yet further CO_2 buildup, giving eventual enhancements to the atmospheric CO_2 in excess of 30% and 15%. One can thus conclude that temperature–pCO_2 feedback is likely to be important on long time scales if a significant fraction of the recoverable fossil fuel resource is used and if the climate sensitivity is high.

Changes in the biological pump

The most likely way in which the biological pump could change in strength in response to a change in climate is if climatic change leads to a change in the rate of mixing between the mixed layer and deep ocean. This would change the rate of supply of nutrients to the mixed layer.

One of the more robust features of coupled atmosphere–ocean climate models is that warming of the climate at rates comparable to that anticipated during the 21st century will induce an at-least-temporary reduction in the intensity of the large-scale, thermohaline overturning in the oceans (Washington and Meehl, 1989; Mikolajewicz *et al.*, 1990; Manabe and Stouffer, 1994; Harvey, 1994). A reduction of 30–50% in the thermohaline overturning by the time CO_2 doubles is a common and quite credible model result.

The possibility of a significant reduction in the intensity of oceanic overturning has prompted a number of studies of the effect of changes in oceanic overturning on atmospheric CO_2 concentration. A change in ocean circulation will affect atmospheric CO_2 by altering the net vertical transfer of DIC associated with the sinking of cold water in polar regions and the upwelling of CO_2-rich water from depth elsewhere; and by altering the availability of nutrients in the surface layer and hence in the rate of biological production and export of particulate carbon from the surface layer. However, the recent results of Maier-Reimer *et al.* (1996) and Sarmiento and Le Quéré (1996) suggest that a 50% reduction in the intensity of thermohaline overturning would reduce the rate of uptake of anthropogenic CO_2 by the oceans by only 10% or less.

7.7 Impulse response for CO_2

Our understanding of the response of the oceans and of the terrestrial biosphere to the injection of CO_2 into the atmosphere, in the absence of climate–carbon cycle feedbacks, can be summarized by running the oceanic and/or terrestrial components of a carbon cycle model and examining what happens when a pulse of CO_2 is suddenly added to the atmosphere. The variation through time in the amount of the injected carbon that remains in the atmosphere is called the *impulse response*.

Figure 7.4 shows the impulse response when the oceans alone absorb the injected CO_2, when the terrestrial biosphere alone absorbs the injected CO_2 (assuming a middle estimate of the CO_2-fertilization effect), and when the oceans and terrestrial biosphere act together to absorb the injected CO_2. These responses were computed using the carbon cycle model of Harvey and Huang (1999) for a sudden injection of 1 Gt C. For these calculations, no dissolution

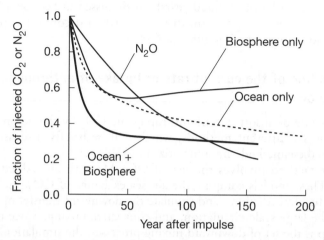

Figure 7.4 The impulse response for CO_2 taking into account removal by the ocean only, by the terrestrial biosphere only, and by the ocean and terrestrial biosphere acting together. Also shown is the impulse response for N_2O. All CO_2 results were computed using the carbon cycle model of Harvey and Huang (1999), while the N_2O impulse corresponds to an N_2O lifetime of 120 years.

of $CaCO_3$ sediments was allowed to occur. The results shown in Figure 7.4 are similar to those obtained by many other models. Focusing first on the ocean-only case, the first 20% of the carbon is removed in less than 10 years. Successive increments take progressively longer to be removed. The final 18% is not removed at all, which is equivalent to giving it an infinitely long time constant. Thus, unlike other GHGs, the removal of a pulse of CO_2 is not characterized by a single time constant. Rather, the time constant for the removal of the remaining CO_2 increases as more CO_2 is removed. When the terrestrial biosphere alone absorbs CO_2, the rate of removal is initially comparable to that due to the ocean alone. However, the biosphere absorption ceases around 40 years after the emission pulse, followed by a weak return flow of carbon to the atmosphere. This is a result of an increase in the respiration flux as respirable carbon accumulates in the soil pools. When the terrestrial biosphere and ocean act together, there is a markedly faster initial decrease in the atmospheric CO_2 pulse then when either act alone, but little difference in the total carbon uptake after 200 years.

Also shown in Figure 7.4 is the rate of decay of a pulse of N_2O, which is governed with a single exponential time constant of 120 years (Table 2.2). Comparison with the impulse response for CO_2 underlines the fact that the removal of CO_2 is governed by multiple time constants, being considerable faster than the 120-year N_2O time constant at the beginning, and being considerable slower after about 200 years. Furthermore, the amount of the N_2O pulse that remains in the atmosphere asymptotically approaches zero, whereas the CO_2 pulse does not do so for all practical purposes (complete removal requiring on the order of 200,000 years). As discussed in Section 11.6, this latter difference creates significant difficulties when it comes to creating a single index for intercomparing different GHGs.

7.8 Estimation of the current rate of uptake of anthropogenic CO_2 by the oceans

The current rate of absorption of anthropogenic CO_2 by the oceans can be estimated by six different methods, which will be briefly discussed here. Results from these methods are summarized in Table 7.1.

The first method involves the use of OGCMs with embedded carbon chemistry. These models compute the air–sea exchange of CO_2, include the important chemical reactions, and simulate the downward transfer of DIC by diffusion, the large-scale circulation, and convective mixing. Since the rate-limiting step is the set of downward mixing processes, the simulation of these processes in an OGCM largely determines its rate of uptake of anthropogenic CO_2. The computed uptake over the period 1980–1989 for three OGCMs and one 2-D ocean model ranges from 1.5 to 2.1 Gt C yr^{-1}. In most cases, it is thought – based on known deficiencies in the models – that the models tend to underestimate the uptake of anthropogenic CO_2.

Table 7.1 Estimates of the rate of uptake of anthropogenic CO_2 by the oceans during the 1980s. See Table 5.1 for definitions of the acronyms GFDL and LODYC

Method	Rate of oceanic uptake (Gt C yr^{-1})	Reference
Uncalibrated 3-D and 2-D models (1980–1989)		
Hamburg OGCM	>1.5	Orr (1993)
LODYC OGCM	2.1	Orr (1993)
GFDL OGCM	>1.9	Sarmiento *et al.* (1992)
2-D model	>2.1	Stocker *et al.* (1994)
1-D models with a priori *calibration using multiple tracer data*		
Box-diffusion model	2.32	Siegenthaler and Sarmiento (1993)
Upwelling–diffusion model	2.15	Siegenthaler and Joos (1992)
Upwelling–diffusion model	1.84	Harvey and Huang (1999), 1980–1989
Direct estimation		
	1.1–2.4	Sarmiento and Sundquist (1992)
Analysis of ^{13}C penetration		
	2.1 ± 0.9	Heimann and Maier-Reimer (1994)
Analysis of O_2 data		
	≈2.0	Keeling *et al.* (1996)
Analysis of atmospheric CO_2 and $^{13}C/^{12}C$ variations		
	1.1	Francey *et al.* (1995), 1982–1992

The second approach involves the 1-D upwelling–diffusion model or some variant of it, with the mixing coefficients previously calibrated to replicate the observed pre-industrial distribution of various oceanic tracers (such as DIC, ^{14}C, and ^{13}C), and the changes in tracer distributions since the start of anthropogenic interference. This approach yields a range of 1.8–2.3 Gt C yr^{-1} for the oceanic uptake during the 1980s.

The third approach is to measure directly the global air–sea CO_2 flux. This flux depends only on the measured mixed layer and atmospheric pCO_2 and on the air–sea exchange coefficient, k_{as}, according to the relation

$$F_{CO_2} = k_{as}((pCO_2)_a - (pCO_2)_{ML}) \qquad (7.2)$$

The observed mixed layer pCO_2 at any given time includes the net effect of downward mixing in the ocean up to that time, so mixing processes therefore do not need to be explicitly considered. The difficulties with this approach are that (i) the mixed layer pCO_2 varies strongly with location and season (particularly at high latitudes), so that a large number of pCO_2 measurements are needed; and (ii) the magnitude of the air–sea exchange coefficient is quite uncertain, and also varies strongly with location and season. Sarmiento and Sundquist (1992) use this approach but, not surprisingly, assigned a rather larger uncertainty to their estimate: 1.1–2.4 Gt C yr^{-1}.

The fourth approach takes advantage of the fact that the release of ^{13}C-depleted carbon from the oxidation of fossil fuels and biomass has caused the atmospheric ^{13}C:^{12}C ratio to decrease during the past two centuries. The penetration of this anomaly into the ocean in principle provides another constraint on oceanic uptake of CO_2. After adjusting uncertain input data within their uncertainty range so as to achieve internal consistency, Heimann and Maier-Reimer (1994) obtained an uptake of 2.2 ± 0.8 Gt C yr^{-1} for the period 1970–1990.

The fifth method utilizes the fact that the absorption of anthropogenic CO_2 by the oceans has no effect on the atmospheric supply of O_2, while absorption by the terrestrial biosphere (through enhanced photosynthesis) produces O_2. If the amount of atmospheric O_2 falls faster (slower) than it is consumed by the burning of fossil fuels, there must be a net terrestrial source (sink) of CO_2 to the atmosphere. Measurement techniques have recently become precise enough to permit determination of trends in the amount of O_2 in the atmosphere, and preliminary results indicate that the northern land biota and global oceans each absorbed about 2.0 Gt C yr^{-1} during 1991–1994. The O_2 data also imply that the tropical land biota were neither a strong source nor a strong net sink of CO_2 (Keeling et al., 1996). Given estimated emissions from tropical deforestation of 1.6 ± 1.0 Gt C yr^{-1} (Schimel et al., 1996), this implies a comparable uptake by undisturbed tropical forests due to the effects of CO_2 fertilization. These results are consistent with the results based on the inversion of atmospheric ^{13}C data, discussed in Section 7.3, which indicate that the tropical biosphere is not a large net source or sink of CO_2. The O_2 concentration can also be measured in air trapped in firn in polar ice caps, and concurrent measurements of the O_2 and CO_2 concentration imply that the terrestrial biosphere was neither a source nor a sink of CO_2 over the period 1977 to 1985 (Battle *et al.*, 1996).

The final approach to estimating the oceanic uptake of anthropogenic CO_2 uses concurrent measurements of global mean atmospheric CO_2 and ^{13}C concentrations. On this basis, Francey *et al.* (1995) estimated an oceanic carbon uptake for 1982–1992 of 1.1 Gt C yr^{-1}.

Based on all the evidence cited above, the average rate of absorption of anthropogenic CO_2 by the oceans during the 1980s is likely to have been 2.0 ± 1.0 Gt C yr^{-1}.

7.9 Atmospheric methane

Compared to CO_2, the cycle of atmospheric methane is simple. The three main processes that need to be considered are (i) the effect of climatic change on natural methane sources (primarily wetlands) and on the rate of removal of CH_4 from the atmosphere; (ii) the potential release of methane from a water–methane compound known as clathrate, which is found in terrestrial permafrost at depths of several hundred metres and in continental slope sediments worldwide; and (iii) the feedback between the methane concentration and its own lifespan. The primary removal process for methane is oxidation with OH to form CO_2, so the decay of atmospheric methane is an additional source of atmospheric CO_2.

Higher temperatures could increase CH_4 emissions from wetlands both by enhancing metabolic rates (Whalen and Reeburgh, 1990) and by enhancing primary productivity (Whiting and Chanton, 1993), but a drop in the water table could lead to decreases in CH_4 production. The computation of methane release from destabilization of methane clathrates requires modelling the downward penetration of heat into oceanic sediments and in permafrost, and doing so in a fairly disaggregated manner (such as on a 1° × 1° latitude–longitude grid). Harvey and Huang (1995) carried out such an exercise but concluded that the potential release of methane from clathrates due to a warmer climate is not likely to be large.

The rate of removal of methane by oxidation with OH depends on the concentration of OH and on the chemical reaction rate, which depends on temperature. The concentration of OH depends on the total emissions of CH_4 (and CO); as emissions increase, the OH concentration decreases and the lifespan of CH_4 increases. However, as the climate warms, the water vapour content in the atmosphere increases, which acts to increase the concentration of OH and thereby reduce the lifespan of CH_4. The greater chemical reaction rate as temperatures increase will also act to reduce the lifespan of CH_4. As noted in Section 3.2, hypothetical temperature-induced changes in CH_4 emissions from wetlands, and temperature effects on the rate of oxidation, can explain the year-to-year variations in the rate of growth in atmospheric CH_4 that were observed during the 1980s and 1990s.

Once the natural methane flux has been estimated, and anthropogenic fluxes added, the atmospheric methane concentration can be modelled as

$$\frac{dC}{dt} = F - \frac{C}{\tau} \tag{7.3}$$

where F is the total methane source flux, and τ is the atmospheric lifespan of methane. It can be updated periodically based on changes in the concentration of methane and in temperature, using the results of 3-D atmospheric chemistry models (as discussed in Section 5.3). However, calculations of future atmospheric CH_4 concentration have generally considered the feedback between CH_4 concentration and τ, but not between climate and τ.

Harvey and Huang (1995) tested the impact of a hypothetical feedback between climate and the natural methane flux, assuming that the natural flux doubles for each 10 K warming. They found that a feedback of this magnitude increased the global radiative forcing and hence temperature response by only 5%. To the extent that higher methane fluxes from wetlands are caused by higher primary productivity rather than warmer temperatures, there would be an associated removal of CO_2 from the atmosphere, so feedbacks involving CO_2 and CH_4 would be partially offsetting. Similarly, conditions that would maximize CO_2 emissions from high latitude peatlands as climate warms (namely, a lower water table) would tend to minimize CH_4 emissions (Gorham, 1991), so that large positive feedbacks involving both CO_2 and CH_4 would not occur simultaneously.

7.10 Other greenhouse gases and aerosols

The atmosphere can be treated as a well-mixed reservoir for other greenhouse gases, and the rate of removal is linearly proportional to the concentration of the gas in the atmosphere. Thus, concentrations of these gases are also governed by Equation (7.3), with no feedback between the concentration of the given gas and its own lifespan. Using Equation (7.3), the concentration at some future time $t_1 + \Delta t$ depends on the concentration at time t_1 and on the current rate of emission. Future concentrations thus depend on the *history* of previous emissions.

In the case of aerosols, the atmospheric lifespan is so short (days) that the current concentration can be regarded as depending only on the current emissions, and not in any way on past emissions. As noted in Section 2.5, about 80% of the emitted SO_2 is converted (oxidized) to SO_4^{2-} under present-day conditions, with about 60% of the conversion done by reaction with H_2O_2 (hydrogen peroxide) and the rest by reaction with OH and O_3. As SO_2 emissions increase, there is proportionately less H_2O_2 available for conversion of SO_2 to SO_4^{2-}, so the fractional increase in the atmospheric SO_4^{2-} loading will be smaller than the fractional increase in SO_2 emissions. For a 50% *decrease* in SO_2 emission, the sulphate loading decreases by 40–43% over the dense source regions (Misra *et al.*, 1989). On a larger scale, the H_2O_2 limitation on the conversion from SO_2 to SO_4^{2-} is presumably smaller (since SO_2 will be less concentrated than over the source regions), so the variation of sulphate loading with SO_2 emissions is presumably even closer to linear. However, if both SO_x (primarily SO_2) and NO_x emissions are reduced by 50%, Misra *et al.* (1989) found the decrease in sulphate loading to be less than 30% in the source regions. This is because, with less NO_x, there is more H_2O_2 available to convert SO_2 to sulphate. Once SO_2 is converted into sulphate aerosol, it is distributed by winds and turbulent mixing, and is removed by rainout and dry deposition.

7.11 Summary

This chapter has highlighted the complexity of processes involved in the response of the carbon cycle to anthropogenic emissions of CO_2, and the differences between the responses of atmospheric CO_2 and other greenhouse gases to anthropogenic emissions. Carbon dioxide differs from other GHGs in that it continuously cycles between a number of "reservoirs" or temporary storage depots (the atmosphere, land plants, soils, ocean water, and ocean sediments). All of the other GHGs (except water vapour, which is not directly influenced by human emissions) are removed by either chemical or photochemical reactions within the atmosphere. For this reason, the rate of removal of non-CO_2 GHGs can be characterized by a single time constant (which, in the case of CH_4, can change over time), and the concentration following a pulse emission will eventually decay to zero. The rate of removal of CO_2, in contrast, cannot be characterized by a single time constant. Rather, different portions of a pulse emission can be thought of as being removed from the atmosphere at different rates, with 10–15% of the emitted carbon requiring on the order of 10,000 years to be removed, and another 10–15% or so requiring on the order of 200,000 years to be removed.

Feedbacks between climate and the carbon cycle have the potential to alter the rate of removal of anthropogenic CO_2 from the atmosphere during the next few hundred to 1000 years. Current knowledge indicates that the potential change in the ocean CO_2 sink due to circulation changes or warmer temperatures is likely to be small, and that the effect of potential methane–wetland and methane–clathrate feedbacks is also small. However, the present terrestrial carbon sink could significantly weaken with continuing fossil fuel emissions, and there is a risk of a transient dieback of forests as climatic zones shift, giving a large CO_2 flux to the atmosphere for a century or longer. There will be a tendency for CO_2 and CH_4 feedbacks involving thawing of permafrost to be negatively correlated; that is, if one is comparatively strong, the other will be comparatively weak, and vice versa. Changes in the strength in ocean overturning also have partly compensating effects on atmospheric CO_2, by altering the rate of upwelling of nutrients and of CO_2-rich deep water. This minimizes the effect of such changes. Terrestrial nutrient feedbacks appear to limit changes in carbon storage due to both temperature enhancement of respiration and CO_2 enhancement of photosynthesis. However, these conclusions must be tempered by the knowledge that the climate system is being moved by human intervention into a state which is unprecedented in recent geological history, and current models could be omitting key processes which, if included, might give a different CO_2 buildup in response to anthropogenic emissions than expected. Scenarios of the expected CO_2 buildup are presented in Chapter 11, but first, we continue our process-based discussion of the climate system response to human emissions of GHGs.

Questions for further thought

1. Why does a higher atmospheric CO_2 concentration have relatively little effect on the rate of photosynthesis in C_4 plants, compared to C_3 plants?
2. Contrast the effect of higher CO_2 on seasonal plant productivity through its effect on photorespiration and on stomatal conductance.
3. Compare and contrast the terrestrial biosphere and oceanic sinks for anthropogenic CO_2 in terms of (a) the processes involved, (b) the rapidity of uptake, (c) the degree of uncertainty, (d) the expected change in the annual rate of uptake over time, and (e) the permanence or reversibility of the carbon storage.

Chapter 8

The regional equilibrium response to a doubling of the atmospheric concentration of CO_2

In this chapter we examine the characteristics and geographical patterns of the climatic changes that are simulated by 3-D AGCMs to result from a doubling of atmospheric CO_2. These climatic changes are computed by first running a model with the present (or near-present) concentration of CO_2 until it has settled into a statistically constant climate, then suddenly increasing the concentration of CO_2 and allowing the simulated climate to reach a new, statistically constant state. In the vast majority of cases, a doubling of the atmospheric CO_2 concentration is imposed. The difference between the two climates is then determined. In some cases a second simulation, without increasing the CO_2 concentration, is continued from the exact point where the CO_2 increase was applied. This simulation is referred to as the "control" simulation, and this is what is compared with the perturbed simulation. The resultant climatic change is often referred to as the "equilibrium" climatic change. In order to achieve a new equilibrium climate after only a few decades of simulated time, only the mixed layer of the ocean is included. If the AGCM were coupled to an OGCM, representing the full depth of the ocean, then about 1000 years of simulated time would be required before the new equilibrium would be achieved. This in turn would require a prohibitively large amount of computer time. Owing to the absence of the deep ocean in a mixed layer-only simulation, the horizontal heat transport by the ocean for the present climate has to be specified, and then held constant as the climate changes.

The use of AGCM-mixed layer models, rather than coupled AOGCMs, introduces two likely sources of error in the simulated geographical patterns of climatic change that can be expected at any given time. First, the geographical patterns of climatic change during the transition from one climate to another could be quite different from equilibrium patterns of climatic change. This is because the *delay* in climatic warming is expected to vary regionally because of differences in the rate of downward mixing of heat into the oceans. The resulting differences in surface warming could be further amplified by differences in cloud feedbacks. This is because the magnitude and even the sign of cloud feedback in a given region seem to depend on the regional patterns of climatic change, and not just on the overall magnitude of climatic change. Second, the geographical patterns of *equilibrium* climatic change itself, as simulated by AGCM-mixed layer models and AOGCMs, are likely to differ. This is due to the fact that changes in the oceanic circulation and associated heat transport can occur in AOGCMs, but these are implicitly held constant in AGCM-mixed layer models. All of these issues will be discussed in the next chapter. Nevertheless, the equilibrium response to a CO_2 doubling, as simulated by AGCM-mixed layer models, serves as a useful benchmark for comparing different AGCMs, for understanding the role of different processes in climatic change, and for gaining an appreciation of the potential magnitude of climatic change at the regional scale.

8.1 Global mean temperature response

Climate sensitivity is defined as the ratio of the global mean change in temperature to the global mean radiative forcing. As discussed in Section 6.6, the climate sensitivity depends on a variety of fast radiative feedback processes. These feedbacks affect the climate sensitivity by adding to or subtracting from the initial radiative perturbation; positive feedbacks add to the initial perturbation, thereby provoking further temperature change, while negative feedbacks subtract from the initial perturbation. In principle, the climate sensitivity can be estimated by measuring the strengths of the major individual feedback processes from observations, or by attempting to simulate these feedback processes in 3-D AGCMs. The problem with the observational approach is that it is rarely possible to directly observe the feedback processes that determine climate sensitivity. The problem with computer models is that they are simplifications of nature, and not nature itself.

The climate sensitivity is believed to be approximately constant for radiative forcing perturbations of up to a few $W\,m^{-2}$, and to be largely independent of the specific combination of forcings that add up to a given total forcing when changes in the concentration of different well-mixed greenhouse gases, in non-absorbing aerosols, and in solar luminosity are combined. For this reason, the climatic response to a CO_2 doubling can be used as an overall indicator of climate sensitivity. Given this climate sensitivity, the global mean temperature response scales up or down with the total forcing.

Estimates based on computer models

Table 8.1 lists the steady-state, global mean change in surface air temperature obtained with coupled AGCM-mixed layer models for a doubling of atmospheric CO_2. In most cases, the global mean warming for a CO_2 doubling is 2.0°C to 4.0°C. The single most important feedback in these models is a positive water vapour feedback, owing to the increase in the amount of water vapour in the atmosphere as the climate warms. The cloud feedback is the net result of a large number of changes in clouds. Table 8.2 lists some of the important cloud changes and their qualitative effect on climate sensitivity (the actual

Table 8.1 Global mean change in surface air temperature (ΔT, K) for a doubling of atmospheric CO_2 as obtained by various AGCMs coupled to a slab ocean

Model	ΔT	Reference
BMRC	2.1	Colman and McAvaney (1995)
Genesis	2.1	Pollard and Thompson (1994)
	2.3	Pollard and Thompson (1994)
UKMO	2.5	Johns *et al.* (1997)
	2.8	Murphy and Mitchell (1995)
GLA	2.6[a]	Sellers *et al.* (1996)
	2.8[b]	Sellers *et al.* (1996)
CCC	3.5	Boer *et al.* (1992)
GFDL	4.0	Manabe *et al.* (1992)
NCAR	4.0	Washington and Meehl (1989)
	4.6	Washington and Meehl (1993)
CSIRO	4.3	Watterson *et al.* (1997)
GISS	4.2[c]	Rind *et al.* (1995)
	4.8[d]	Rind *et al.* (1995)

Notes: The Genesis model is an outgrowth of the NCAR AGCM. See Table 5.1 for definitions of the acronyms used here.

[a] Standard model version.

[b] Version with reduced stomatal conductance in response to a higher atmospheric CO_2 concentration. This limits evaporation, causing greater surface and near-surface warming over land.

[c] Standard model version, having sea ice that is too thick for the present climate.

[d] Model version in which sea ice thickness for the present climate matches observational estimates.

cloud changes in any given AGCM can be different from those shown in Table 8.2). The net effect of cloud feedbacks in most models is small to slightly negative, although a sizeable net positive feedback occurs in some models. The retreat of seasonal snowcover and sea ice as climate warms also serves as a positive feedback, since it allows more solar radiation to be absorbed.

Direct observation of the water vapour feedback

There is no doubt that the total amount of water vapour in the atmosphere will increase as the climate warms. This follows from very fundamental principles involving the fact that the driving force for evaporation from the ocean – the difference between surface and atmospheric vapour pressures – tends to increase as the surface temperature increases, combined with the fact that the atmosphere's ability to hold water increases as the air warms. As summarized in Section 4.4, observations made during recent decades indicate that the atmospheric water vapour content has indeed increased as the climate warmed.

Since water vapour in the upper troposphere is particularly effective in trapping heat, increases in this region are particularly important to climate sensitivity (Section 6.1). As discussed in Section 4.4, the available rawinsonde observations are not reliable for determining long-term trends in the amount

Table 8.2 Qualitative effect of various changes in cloud characteristics that could occur as the climate changes. Also given are references where the indicated effect is quantitatively assessed or included in climate simulations. The actual changes in cloud properties that occur as the climate warms may be different from those indicated here, and sometimes differ from model to model

Cloud property change	Effect on climate	Example reference
Decrease in amount of low cloud	Warming	Le Treut and Li (1991)
Increase in amount of high cloud	Warming[a]	Wetherald and Manabe (1988)
Increase in cloud height	Warming	Mitchell and Ingram (1992)
Increase in water content of stratus clouds	Cooling	Somerville and Remer (1984)
Increase in water content of cirrus clouds	Warming or cooling	Lohmann and Roeckner (1995)
Increase in ratio of water droplets to ice crystals	Cooling through enhanced reflectivity	Senior and Mitchell (1993)
	Warming through increased particle fall velocities	Senior and Mitchell (1993)

[a] This assumes that high clouds have a net heating effect. However, Del Genio *et al.* (1996) calculate a net cooling effect for high tropical cirrus clouds in the GISS AGCM.

of water vapour in the upper troposphere. However, the available datasets are adequate for assessing the spatial distribution of interannual changes in water vapour and their correlation with interannual changes in surface temperature. Much of the interannual variability is related to the El Niño oscillation, which involves large-scale oscillations in surface temperatures in the tropical Pacific Ocean and a shift in the main region of convection between Indonesia and the eastern Pacific Ocean. Since the changes in surface temperature are associated with major shifts in the tropical circulation that are unique to the El Niño oscillation, the *local* correlation between temperature and humidity is obviously not going to be applicable to long-term, large-scale climatic change (the same is also true for seasonal changes in temperature and humidity).

Sun and Oort (1995) have attempted to circumvent this problem by examining the correlation between atmospheric water vapour and surface temperature averaged over the *entire* tropics (30°S–30°N). Based on data for the period from May 1963 to December 1989, they found that the amount of water vapour in the tropical atmosphere increased with increasing temperature as follows:

- by 70–75% of the increase that would occur with fixed RH in the 1000–900 mb layer;
- by only 15% of the increase expected with fixed RH at 700 mb; and
- by 70% of the increase expected with fixed RH at 300 mb.

In the planetary boundary layer (1000–900 mb), the change in water vapour is most directly tied to the thermodynamically driven increase in the rate of evaporation. There is also a sizeable relative increase in the upper troposphere, related no doubt to an increase in the rate of detrainment of moisture from the tops of cumulus cloud columns. However, in the middle troposphere there is almost no increase in the amount of water vapour, which can be explained by an increase in the rate of subsidence of relatively dry air from the upper troposphere (even though this air becomes moister, it is still much drier than the air below).

Since AGCMs simulate close to constant RH at all heights as the climate warms, the results of Sun and Oort (1995) – assuming that they are applicable to long-term climatic change – imply that the water vapour feedback is too strong in AGCMs. However, the potential error in the strength of the global mean water vapour feedback is no more than 10%.

An alternative to examining the correlation between changes in surface temperature and atmospheric water vapour is to examine the relationship between specific processes and atmospheric water vapour. Soden and Fu (1995) examined the relationship between deep convection and upper tropospheric water vapour as inferred from satellite data. They found that regions of deep convection are associated with greater moisture in the upper troposphere in the tropics. The more important questions, however, are (i) how the *aerially averaged* humidity changes through *time* as the aerially averaged frequency of convection changes, and (ii) how climatic change will affect convection. Currently available data cannot reliably answer these questions. However, Soden (1997) showed that current AGCMs can reproduce many of the observed year-to-year variations in the strength of greenhouse gas trapping of radiation if driven by observed variations in SST. This indicates that processes involving water vapour are reliably simulated – the same processes that produce a positive longterm water vapour feedback.

With regard to middle and high latitudes, Del Genio *et al.* (1994) find that eddies and mean meridional motions are the dominant factors influencing upper tropospheric moisture outside the tropics. The moistening effect of eddies and mean meridional motions would both increase as the climate warms.

The evidence reviewed above supports the conventional wisdom that the water vapour feedback is positive. Observations indicate that total precipitable water increases as the climate warms, which will exert a positive feedback. Observations of interannual variability indicate that the amount of water vapour also increases in the upper troposphere, although the increase in the middle troposphere is probably less than simulated by current AGCMs. Observational data also indicate that tropical greenhouse gas trapping increases as sea surface temperature increases. However, the strength of the longterm temperature-water vapour feedback likely depends in part on the *patterns* of surface temperature change, since convection and its effects depend on both spatial patterns of warming and the average warming.

Thus, there is no reason to suspect that the water vapour feedback is other than positive, and of a magnitude sufficient to increase the climate sensitivity by at least 50% compared to the case with no feedbacks. Recall (from Section 6.4) that the climate sensitivity for a CO_2 doubling in the absence of any feedbacks is about 1.0°C. The water vapour feedback probably increases this to at least 1.5°C. Negative cloud feedbacks could reduce the climate sensitivity below 1.5°C, but ice and snow feedbacks would increase the sensitivity. Thus, we can accept 1.5°C as a likely lower limit for the climate sensitivity to a CO_2 doubling on the basis of the evidence reviewed so far.

Assessments based on paleoclimatic data

Since climate sensitivity is defined as the ratio of the global mean change in temperature to the global mean radiative forcing, it can be calculated empirically if these two quantities can be estimated for past climates. Hoffert and Covey (1992) did this using data for the Cretaceous Period, when the climate was substantially warmer than at present, and for the peak of the last ice age, when the climate was substantially colder. The climate sensitivity calculated in this way implies a climatic response for CO_2 doubling of 2.3 ± 0.9°C, which falls within and below the lower half of the 2.1–4.8°C range derived from recent AGCMs (Table 8.1). Interestingly, Hoffert and Covey (1992) deduced a similar sensitivity for a climatic warming (the Cretaceous Period) as for a climatic cooling (the last ice age).

Assessments based on the analysis of historical temperature changes

The analysis of global mean temperature changes during the 20th century could, in principle, provide an independent estimate of climate sensitivity. However, because of uncertainty in the magnitude of cooling effects associated with biomass and sulphur aerosols, the observed changes in global mean temperature cannot provide a meaningful constraint on climate sensitivity. This is compounded by uncertainty in the radiative heating or cooling effects associated with changes in O_3 and in the magnitude of solar radiation variations during the last 100 years. Hence, we must first deduce the relative importance of different forcing mechanisms. Constraining the aerosol forcing requires examining the spatial patterns of climatic change. We must also take into account the effect of the oceans in delaying the temperature response to radiative forcings. This problem is discussed more fully in Chapter 9, but it suffices to state here that the observed temperature changes during the past century are consistent with a climate responsiveness for a doubling of CO_2 of 1.0°C to perhaps 3.0°C *if* the effective climate sensitivity during a time-dependent change is the same as the equilibrium sensitivity. This is an assumption that will be fully addressed in Chapter 9.

Synthesis

The first-order response of the climate to a doubling of atmospheric CO_2 is the global mean warming that occurs in equilibrium. Simulations with computer climate models, analysis of paleoclimatic data, and analysis of historical temperature changes indicate temperature changes of 2.1–4.8°C, 1.4–3.2°C, and 1–3°C, respectively. These results are summarized in Figure 8.1.

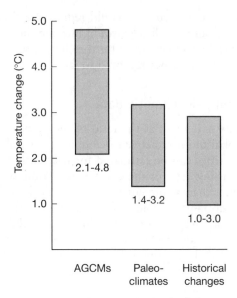

Figure 8.1 The global mean equilibrium climatic change ("climate sensitivity") for a doubling of atmospheric CO_2 as given by AGCM simulations, paleoclimatic data, and historical temperature variations.

8.2 Regional temperature response patterns

As discussed above, the majority of recent simulations with AGCMs give a global and annual mean surface air warming for a doubling of CO_2 of 2.0–4.0°C in equilibrium. However, a single number – the global average temperature change – masks substantial geographical and seasonal differences in the simulated temperature response. The main features of the simulated response patterns, that occur in all models to some extent, are as follows:

- a greater mean annual warming at high latitudes than at low latitudes at the surface and in the lower troposphere, also referred to as a *polar amplification* of the warming;
- a greater warming in winter than in summer at high latitudes, particularly over the oceans;
- greater warming in summer than in winter in those land areas where soils become sufficiently dry in the summer;
- a greater warming in the tropics than at middle and high latitudes in the upper troposphere; and
- a cooling of the stratosphere.

These characteristics of the response of AGCMs are illustrated in Figures 8.2 and 8.3. Figure 8.2 shows the geographical pattern of surface air warming for December–January–February (DJF) and June–July–August (JJA) as simulated by three different AGCMs that have low, intermediate, or high climate sensitivity (the corresponding zonally averaged changes are compared with a larger set of models in Plate 7). Figure 8.3 shows the variation with latitude and height of the zonally averaged mean annual warming as simulated by representative AGCMs using two different parameterizations of convection.

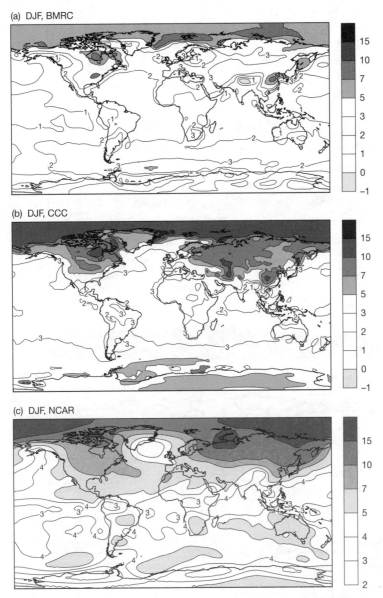

Figure 8.2 Equilibrium change in surface air temperature (°C) following a CO_2 doubling, as simulated by three different AGCMs having low (BMRC) intermediate (CCC) and high (NCAR) climate sensitivity, and averaged over the months of December-January-February (DJF) or June-July-August (JJA). All results are based on data files kindly proved by one of the authors of the following publications, which also describe the model simulations: BMRC, Colman and McAvaney (1995); CCC, Boer et al. (1992); NCAR, Washington and Meehl (1993).

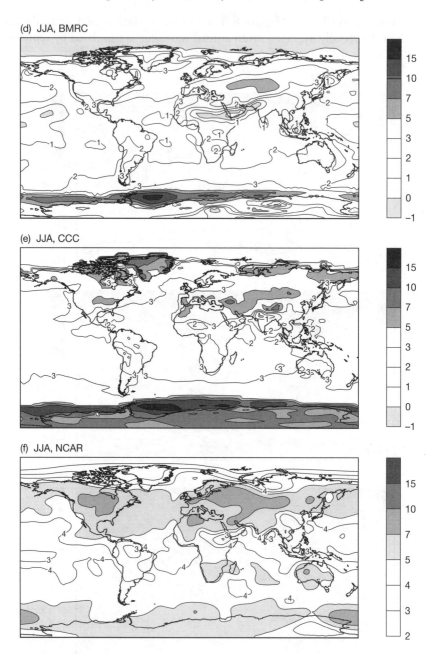

Figure 8.2 Continued

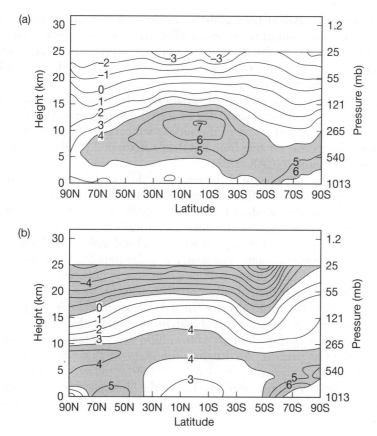

Figure 8.3 Variation with latitude and height of the mean annual, zonally averaged change in temperature in response to a doubling of atmospheric CO_2 as simulated by representative AGCMs using (a) penetrative convection and (b) a moist convective adjustment. Reproduced from Mitchell *et al.* (1990).

Polar amplification of the surface response

The greater surface and near-surface warming at high latitudes is a result of (i) the feedback between temperature and the extent of snow and ice, and (ii) differences in the feedback involving atmosphere lapse rate. As temperatures warm, the extent of ice and snow decreases. This results in a reduction in the surface albedo and an increase in the absorption of solar energy, which tends to amplify the local temperature response. Since ice and snow cover are restricted to high latitudes, the high-latitude temperature response is amplified. The other factor is the lapse rate feedback. At high latitudes the lapse rate tends to increase due to the stability of the high-latitude atmosphere, which restricts the warming to a shallow layer of the atmosphere. At low latitudes the average lapse rate is closely linked to the moist adiabatic lapse rate, which decreases as the climate warms. As discussed in Section 6.5, an increase in lapse rate as surface temperature warms acts as a positive feedback, while a decrease in lapse rate serves as a negative feedback.

Plate 7 compares the latitudinal variation in the zonally (east–west) averaged surface-air warming as simulated by several different AGCMs. All models produce a polar amplification of the surface temperature response relative to the tropical response, ranging from less than a factor of 2 (LMD) to a factor of 3 (CSIRO) or more (NCAR–2).

An important consequence of the polar amplification of the surface temperature response is that the difference between polar and equatorial temperatures decreases (that is, the meridional temperature gradient decreases). As explained in Section 8.3, differences in the polar amplification have important consequences for the simulated changes in rainfall and soil moisture in both tropical and middle latitude locations. The degree of polar amplification is also important to potential changes in mid-latitude storminess, as discussed in Section 8.5.

All climate models predict substantially smaller polar amplification than implied by the analysis of paleoclimatic data. Russian geologists in particular have attempted to reconstruct spatial patterns of climatic change that occurred at various times during the past 100 million years when the climate was warmer than at present. An English summary of this work is found in Borzenkova (1992). Kheshgi and Lapenis (1996) estimated error bars in the latitudinal profile of temperature change for the Holocene Climatic Optimum (5300–6200 years before present). Figure 8.4 shows the latitudinal tempera-

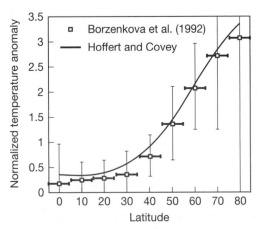

Figure 8.4 The dependence of zonal-mean temperature changes on latitude. Shown are the normalized temperature response profile for the Holocene, as given by Borzenkova (1992) with error bars estimated by Kheshgi and Lapenis (1996), and Hoffert and Covey's (1992) fit to the latitudinal variation in temperature change between the present and the Last Glacial Maximum, and between the present and the Late Cretaceous. Reproduced from Kheshgi and Lapenis (1996).

ture profile for the Holocene, with error bars, along with the latitudinal profile deduced by Hoffert and Covey (1992) for the Last Glacial Maximum and the Late Cretaceous. The middle estimate of the polar amplification is about a factor of 6 – substantially larger than the factor of 2–3 obtained by current climate models (as shown in Plate 7). At present it is not possible to say which approach is more likely to be correct.

Greater warming in winter than in summer at high latitudes

As long as ice or snow is present at the surface, the surface temperature cannot rise above 0°C. Additional heating of the surface will go into melting of ice and snow rather than raising the surface temperature. Where sea ice is thick enough to persist right through the summer in spite of the increased melting, the surface air warming will be strongly suppressed. However, as air temperatures begin to drop during autumn and winter, the thinner sea ice will permit a greater conductive flux of heat from the underlying ocean water (which cannot drop below the freezing point of sea water, about –1.8°C) to the atmosphere. This will reduce the normal seasonal cooling of the atmosphere, with the result that late autumn and winter temperatures in a doubled CO_2 experiment can be more than 10°C warmer than for the control climate (Figure 8.2(b,c)). The greater heat flux to the atmosphere is offset by a greater rate of freezing at the ice base than in the control climate, which releases the latent heat that ends up in the atmosphere. In effect, greater melting of sea ice in summer is offset by greater freezing of sea water in winter, so that the energy that went into the enhanced summer melting is released in winter. This suppresses the summer warming and enhances the winter warming.

Greater warming in summer where soils become drier

When soils become drier, there is less evaporative cooling, which tends to amplify the surface warming. Since soils are projected to become drier in summer over broad regions (Section 8.3), the result can be to cause the largest seasonal warming to occur in summer in some regions. Drier soils would be less effective in creating warmer surface temperatures in winter, should soils be drier then, because there is less energy available to heat the surface in winter.

Greater warming at low latitudes in the upper troposphere

The greater warming of the upper troposphere at low latitudes is a consequence of the increase in latent heating by the condensation of water vapour, which in turn is a result of an increase in the upward pumping of water vapour by cumulus convection as the climate warms. The increase in warming with increasing height in the tropics has two, and possibly three, important consequences. First, it represents a decrease in the lapse rate, which serves as a negative feedback on surface temperature warming (see Section 6.5). Second, it tends to make the middle and upper tropospheric warming at low latitudes greater than at high latitudes. This in turn will increase the meridional temperature gradient in the upper troposphere, which is opposite to the decrease in meridional temperature gradient that occurs in the lower troposphere. As discussed in Section 8.5, the change in storminess in mid-latitude regions depends to a large extent on how the meridional temperature gradient changes, so the changes in the upper and lower troposphere have opposing effects. Third, the increase in the meridional temperature gradient in the upper troposphere is likely to reduce the ability of planetary waves to transport energy upward into the stratosphere, thereby contributing to a cooling of the stratosphere as the surface climate warms. This topic is discussed next.

Cooling of the stratosphere

The change in stratospheric temperature in response to an increase in CO_2 (and other GHGs) is governed by two factors: changes in the *radiative* energy balance, and changes in the *dynamical* heat transports. The latter involves the reduction in upward heat transport associated with planetary-scale waves that is expected to occur if the upper troposphere warms more at low latitudes than at high latitudes (Shindell *et al.*, 1998). This pattern of warming is driven by overall warming itself and the associated increase in tropical convection. Thus, to the extent that dynamical heating of the stratosphere does decrease as the climate warms, the increase in all GHGs will contribute to a cooling of the stratosphere. However, the dynamically induced changes in stratospheric temperature are uncertain.

The effect on the stratospheric radiative energy balance of an increase in GHG concentrations is quite different for different GHGs, owing to differences in the extent to which the gases already absorb radiation. An increase in CO_2 tends to cool the stratosphere, but increases in other GHGs either have very little effect (CH_4 and N_2O) or tend to warm the stratosphere (SF_6 and the halocarbons; see Table 6.1).

According to calculations by Ramaswamy *et al.* (1996), the heating effect of increases in SF6 and the CFCs offsets just over half of the CO_2 cooling effect over the period 1765–1990, but offsets all of the cooling effect over the period 1979–1990. Whether or not there is a net radiative cooling effect in the future depends on the relative increases in CO_2, SF_6, and the halocarbons. The cooling that has occurred during the last 20 years (see Section 4.1) has been overwhelmingly caused by the concurrent decrease in stratospheric ozone, but this effect will diminish as the amount of ozone gradually recovers during the course of the 21st century (WMO, 1998).

The net cooling of the stratosphere, to the extent that it occurs and influences the temperature changes just below the tropopause, is likely to influence how peak tropical cyclone intensity changes as the climate warms, as discussed below in Section 8.5. Stratospheric cooling, whether caused by increases in

GHG concentrations or by the depletion of stratospheric O_3, is also expected to delay the recovery of stratospheric O_3 (Shindell *et al.*, 1998).

8.3 Soil moisture response

The changes in soil moisture that occur as the climate warms are among the most important characteristics of climatic change. If present soil moisture levels can be maintained as the climate warms, the impacts of climatic change on natural ecosystems and agriculture will be much less severe or, in many instances, decidedly positive. If, on the other hand, soils become drier as the climate warms, than the prospect of a warmer climate is much more serious. AGCMs project a general drying of soils as the climate becomes warmer, but disagree sharply concerning the extent of drying and concerning the geographical patterns of soil moisture change. Paleoclimatic data from climates warmer than at present, in contrast, have been interpreted as implying that a warmer climate will be associated with moister soils.

Results from AGCM simulations

The major changes that are simulated by AGCMs to result from a doubling of atmospheric CO_2 are as follows:

- an increase in global mean evaporation from the oceans by 5–10% in most models and a corresponding increase in the global mean rate of precipitation, so that there is an overall intensification of the hydrological cycle;
- a poleward shift in the mid-latitude belt of high rainfall associated with the westerly jet stream;
- an increase of precipitation in mid-latitudes in winter and increases or decreases in summer;
- a tendency for soils to become drier in summer in mid-latitudes;
- a tendency for the Asian and West African monsoons to become stronger; and
- a mixture of increases and decreases in soil moisture in summer in tropical regions.

The reasons for some of these responses are outlined below.

A general tendency for drying of mid–latitude soils in summer is expected based on very fundamental principles, and so can be regarded as a reliable model result (Rind *et al.*, 1990)

Drought tends to occur in nature and in climate models when potential evapotranspiration exceeds precipitation. Potential evapotranspiration (E_p) is the evapotranspiration rate that occurs when water is freely available. It is driven by the difference between the saturation vapour pressure evaluated at the surface temperature, $e_{sat}(T_s)$, and the atmospheric vapour pressure (e_a). Actual evapotranspiration is equal to E_p times some resistance factor that depends on how depleted the soil moisture is. The relationship between saturation vapour pressure and temperature is given by the Clausius–Clapeyron equation, and is

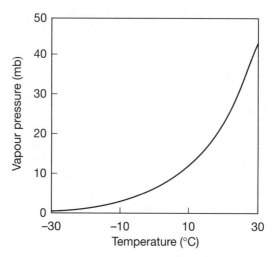

Figure 8.5 The relationship between temperature and saturation water vapour pressure, known as the Clausius–Clapeyron relation.

shown in Figure 8.5. The relationship is highly non-linear, with a strongly concave upward shape. The non-linearity of the Clausius–Clapeyron equation is absolutely crucial to the response of the hydrological cycle and of soils to a warmer climate.

First, with equal warming of the land surface and of surface air over land, $e_{sat}(T_s)$ will increase faster than $e_{sat}(T_a)$ because T_s is warmer than T_a by 1–2°C. In the version of the GISS model analysed by Rind *et al.* (1990), both the land surface and surface air temperature over land warmed by about 4.5°C, which increased $e_{sat}(T_s) - e_{sat}(T_a)$ by 40%. If the surface-air relative humidity is constant (as is roughly the case in AGCMs), e_a will increase even more slowly than $e_{sat}(T_a)$ since $e_a = RH\, e_{sat}(T_a)$, so that $\Delta e_a = RH\, \Delta e_{sat}(T_a)$. In this case, $e_{sat}(T_s) - e_a$ will grow even more quickly as temperatures increase. Thus, E_p over land tends to increase. Two factors (not related to the non-linearity of the Clausius–Clapeyron equation) cause the increase in ocean evaporation to be much smaller than the increase in E_p over land. First, the land surface warms more than the ocean surface, so the increase in $e_{sat}(T_s)$ over the ocean is less than that over land. Second, air temperature over land tends to warm more than the ocean surface, and because this air is advected over the ocean, the air over the ocean warms more than the ocean surface, on average. This allows ocean air to hold more moisture, thereby causing a larger increase in e_a and inhibiting the increase in $e_{sat}(T_s) - e_a$ over the ocean. It also increases the stability of the atmosphere next to the surface, which reduces turbulent mixing and hence evaporation. The greater warming of the land than of the ocean surface, and of oceanic air than of the ocean surface, are consistent and intuitively reasonable results of climate models. The net result is that, in the GISS model, evaporation from the ocean (and global mean precipitation) increases by 11% on average while E_p from land increases by 30%. As result, E_p from land tends to increase faster than precipitation over land, and soils tend to become drier on average.

The overall tendency for drier soils is exacerbated at low and middle latitudes by further effects related to the non-linearity of the Clausius–Clapeyron equation. Low and middle latitudes see the largest increase in E_p because they are warmer to begin with. In contrast, the largest increase in precipitation will tend to occur where the air is coldest and hence easiest to saturate (i.e., at high latitudes). Thus, increases in E_p will have a particularly strong tendency to exceed increases in precipitation at low and middle latitudes.

Changes in both mid–latitude and low–latitude precipitation depend on the polar amplification of the temperature response (Rind, 1987, 1988)

The greater the polar amplification of warming, the greater the decrease in the meridional temperature gradient that occurs and the greater the decrease in mid-latitude synoptic-scale (10^3 km) storms. This leads to a reduction in water supply at mid-latitudes compared to the case with smaller polar amplification. For a given global mean warming, a greater polar amplification implies a smaller tropical warming and evaporation increase, which also decreases rainfall in mid-latitudes. The pattern of rainfall increases and decreases at low latitudes also depends on the degree of polar amplification, as this affects where increases or decreases in subsidence and rising motion occur.

The simulation of soil moisture conditions in late winter and early spring for the present climate is critical to the simulated change in summer soil moisture when CO_2 is increased

Meehl and Washington (1988) analysed the change in summer soil moisture for CO_2 doubling as predicted by the NCAR and GFDL models. The GFDL model generally simulated mid- to high-latitude soils as being saturated in winter at present, while the NCAR model simulated the soils as being unsaturated in winter. Both models predict an increase of winter precipitation, which is lost as runoff in the GFDL model but leads to greater spring soil moisture in the NCAR model. Both models undergo about the same increase in spring-to-summer soil moisture loss for a doubled CO_2 climate. However, soils in the NCAR model receive a boost in their moisture content in early spring under a warmer climate, so the NCAR model projects little to no drying of soils in summer compared to present, while the GFDL model projects significant drying. On the other hand, the NCAR model simulates close to zero present-day summer soil moisture in mid-latitudes, such that there is very little scope for further summer soil moisture decreases. The correct soil moisture change is therefore likely to lie somewhere between the NCAR and GFDL model results.

The treatment of snowmelt also affects the change in summer soil moisture through its effect on spring soil moisture in the unperturbed simulation. Mitchell and Warrilow (1987) compared the effect on summer soil moisture of a CO_2 doubling using two versions of the UKMO GCM: in one, all snowmelt infiltrates the soil unless the soil is saturated, while in the second all snowmelt is assumed to produce runoff if either of the middle two soil layers is subfreezing. In the first case, mid-latitude soil conditions for present CO$_2$ are simulated to be saturated in spring. The increase in winter precipitation and subsequent snowmelt under doubled CO$_2$ therefore does not increase initial soil water amounts prior to the period of enhanced evaporation in late spring and summer, so late summer soil moisture decreases when CO$_2$ is increased. In the second case, soils are not simulated to be saturated in spring for present CO$_2$, so that the winter precipitation increase under doubled CO$_2$ leads to greater spring soil moisture. This increase in spring moisture largely offsets the increased spring–summer evaporation, resulting in only a minor decrease in soil moisture by late summer. These two parameterizations represent extreme cases, but they demonstrate the importance of snowmelt modelling through its control on the late spring soil moisture in the unperturbed climate.

Earlier snowmelt and an earlier onset of spring rains also contribute to decreases in summer soil moisture
Wetherald and Manabe (1995) find that these changes are the main reasons for a decrease in high-latitude soil moisture in summer, and are a contributing factor to the decrease in mid-latitude soil moisture. However, these effects will be mitigated to the extent that they lead to greater end-of-spring soil moisture.

Synthesis of AGCM results

Although AGCMs simulate widespread decreases in summer soil moisture, there are significant regional variations in the magnitude of the drying as simulated by any one model, and significant disagreements at the regional level between different models. The quantitative soil moisture changes simulated by current AGCMs are of low reliability. A good simulation of present amounts of ice and snow, precipitation patterns, and soil moisture is a *necessary* (but not *sufficient*) condition for an accurate simulation of the changes in soil moisture associated with a warmer climate. However, basic principles that are independent of the details of any given model provide several reasons for expecting an overall decrease in summer soil moisture at low and middle latitudes, and for expecting the decreases to become larger and more widespread the more the climate warms.

Implications from the study of past climates

In contrast to computer-based results, analyses of inferred moisture patterns associated with warm intervals in the geological past suggest that continents will become moister rather than drier as the climate warms. However, past warm climates are not suitable guides to the moisture conditions that would occur in association with future warming for at least two reasons: (i) as discussed by Crowley (1990), the seasonal solar radiation pattern (in the case of the mid-Holocene and last-interglacial warm periods) or polar ice cover (in the case of earlier warm periods) were quite different from that which will be associated with warming during the next few centuries, and these differences could readily produce a substantially different precipitation response to warming; and (ii) a significant part of the moisture response to past warm intervals probably involved gradual adjustments in soil properties that would not be involved in moisture changes during the 21st century (Lapenis and Shabalova, 1994). In particular, as climate warms and precipitation in coastal regions increases, the soil moisture-holding capacity is believed to increase. This increase allows for greater storage of rainfall after rainfall events, greater re-evaporation of the moisture, and greater precipitation further inland. However, the adjustment in soil moisture-holding capacity would occur over a period of several hundred years. Hence, data from past climates cannot be taken as evidence that a warmer world would lead to moister continents during the next century. Rather, simple principles indicate that, for fixed soil properties, increasing warming should lead to an increasing tendency for drying in continental interiors, as discussed above.

8.4 Changes in the frequency of days with heavy rain

AGCMs generally show an increase in the proportion of days with heavy rainfall as the climate warms, particularly in the tropics (Kattenberg et al., 1996, Section 6.5.6). Hennessy et al. (1997) found this to be the case even in places where the total number of rainfall days decreases. Gordon et al. (1992) found the increase in the intensity of daily precipitation to be a much clearer response of their model to warming than the changes in the total amount of precipitation. An increase in the intensity of rainfall will contribute to the widespread reduction in soil moisture obtained by AGCMs, since relatively more of the rainfall will be lost as runoff.

8.5 Changes in mid-latitude storms

Some AGCMs give a decrease in the strength and/or frequency of mid-latitude storms (e.g., Branscome and Gutowski, 1992; Stephenson and Held, 1993; Zhang and Wang, 1997), while others give an overall increase in storminess (e.g., Hall et al., 1994). These contradictory results arise because mid-latitude storminess depends on a number of factors which change in opposite ways as the climate warms, so that the net effect is uncertain. The meridional gradient of surface temperature tends to decrease (tending to reduce the jet stream speed and associated storminess), but the meridional gradient of upper tropospheric temperature tends to increase (which tends to increase storminess). Changes in the vertical temperature gradient (or lapse rate) also affect the occurrence of extratropical cyclones. Regional changes in the temperature gradient can be quite different from the zonally averaged change. For example, in the AGCM-mixed layer simulation of Hall et al. (1994), only slight warming occurs over southern Greenland, but strong warming occurs over western Europe. This increases the temperature gradient in the north-east Atlantic, and leads to a marked intensification and downstream shift in the Atlantic storm track. Stephenson and Held (1993), in contrast, obtain a weakening of the Atlantic storm track but no change in the Pacific storm track.

8.6 Changes in El Niños, monsoons, and tropical cyclones

Three of the most important features affecting weather and climatic variability in the tropics are the El Niño oscillation; the Australian, Asian, and west African monsoons; and tropical cyclones. The term El Niño refers to the periodic occurrence of warmer than average sea surface temperatures (SSTs) in the eastern equatorial Pacific Ocean. It is associated with colder than average SSTs in the western equatorial Pacific Ocean, greatly increased rainfall in the eastern equatorial Pacific and the adjacent parts of South America, and reduced rainfall in the western equatorial Pacific, south-east Asia, and northern Australia. Altered weather patterns occur in parts of the northern hemisphere mid-latitudes (Fraedrich *et al.*, 1992). The opposing temperature pattern (cold eastern Pacific and warmer than usual western Pacific) is referred to as La Niña. The El Niño drives a rising motion over the eastern Pacific Ocean and compensating descending motion over the western Pacific Ocean, which explains the rainfall anomalies in these regions that were noted above. Variations in the strength of the monsoons and in the occurrence of tropical cyclones are associated with El Niño–La Niña fluctuations. There are additional, direct links between the strength of the Asian monsoon and the occurrence of cyclones in the western Pacific and Indian Oceans. It is therefore appropriate to discuss the projected changes in these phenomena together, as they form an interlinked system.

Monsoons

The monsoons are driven by differences in the temperature contrast between continents and oceans. Moisture-laden surface air blows from the Indian Ocean to the Asian continent, and from the Atlantic Ocean into west Africa during the NH summer, when the land masses become much warmer than the adjacent ocean. Monsoon-like winds also blow from the Gulf of Mexico onto the south-central United States, and onto northern Australia during the SH summer. In winter, as the continents become colder than the adjacent ocean, higher surface pressure develops over the continents and the surface winds blow towards the ocean.

Climate models indicate an increase in the strength of the summer monsoons in response to increasing CO_2. This is a credible result, as there are three very simple reasons why it should occur:

- The continents tend to warm more than the oceans, especially in summer. This increases the land–sea temperature contrast, which is the ultimate driver of the summer monsoon.
- In the case of the Asian monsoon, decreased snowcover on the Tibetan plateau – which can be expected to occur as the climate warms – has been

Table 8.3 Changes in the characteristics of the Asian monsoon associated with a doubling of atmospheric CO_2 concentration, based on AGCM simulations by Meehl and Washington (1993), Bhaskaran et al. (1995), and Hirakuchi and Giorgi (1995)

Monsoon characteristic	Change
Intensity of air flow	Increases in some areas
Position	Shifts north over India
Onset date	No detectable change
Seasonal mean precipitation	Generally increases by 10–20%
Interannual variability of mean seasonal precipitation	Increases by 25–100%
Proportion of days with heavy rainfall	Dramatic increase

    observed today to be associated with a stronger summer monsoon (Dey and Bhanu Kumar, 1982; Dickson, 1984).

- With a warmer climate, the atmosphere can hold more water vapour, so that winds converging over the landmass will be able to supply more moisture.

The first two points pertain to the vigour of the summer monsoon circulation, while the third point pertains to the moisture transport associated with a given air flow. The changes in a number of monsoon variables, as simulated by a variety of models, are given in Table 8.3.

Two factors might counteract the expected tendency for the strength of the monsoons to increase, at least in some regions: (i) both the Asian and Australian monsoons tend to be weaker during El Niños (Kiladis and Diaz, 1989; Evans and Allan, 1992; Kane, 1997), so if there is an increase in the occurrence or strength of El Niños, this could reduce the average strength of the monsoons; and (ii) the cooling effect of anthropogenic sulphate aerosols over the Asian landmass would tend to reduce the strength of the monsoon by reducing the land–sea temperature difference in summer (Meehl et al., 1996; Mitchell and Johns, 1997).

El Niños

Simulations with coupled AOGCMs, or with AGCMs using prescribed background SSTs and El Niño-like SST anomalies, indicate the following effects of a warmer climate:

- no change in the frequency of El Niños (Meehl et al., 1993; Knutson et al., 1997);
- greater differences in precipitation and soil moisture between El Niño and La Niña years (Meehl et al., 1993), no change in interannual variability of SST (Tett, 1995) or of rainfall (Smith et al., 1997), or a decrease in interannual variability of SST (Knutson et al., 1997); and
- very different patterns of change in sea level and upper tropospheric pressure outside the tropics in response to El Niños, compared to the changes

that occur under the present climate (Meehl *et al.*, 1993).

The conflicting results concerning changes in wet and dry extremes are due to several competing factors acting at once. The first factor arises from the non-linearity of the Clausius–Clapeyron equation. A given change in SST will have a larger effect on evaporation the warmer the base climate. The greater increase in evaporation in the eastern Pacific when an El Niño occurs in a warmer climate will lead to a greater increase in precipitation in this region, greater rising motion, and greater compensating subsidence in the western Pacific. Similar changes, but with the opposite geographical pattern, will occur during La Niña events. Hence, both wet and dry extremes will tend to be intensified as the climate warms. However, the increase in atmospheric stability at low latitudes that occurs in a warmer climate (owing to a decrease in the lapse rate) will tend to dampen the effects of the non-linear increase in evaporation. This was sufficient to neutralize the tendency for increased variability in the simulations of Smith *et al.* (1997).

Changes in the average difference between SSTs in the cool eastern tropical Pacific and the warm western tropical Pacific Ocean will also influence the extent to which the amplitude of El Niño anomalies changes. However, there is no agreement at present as to how this gradient will change as the climate warms.

Tropical cyclones

There are a number of largely independent issues concerning the effect of a warmer climate on tropical cyclones (TCs). The first issue is the global area affected by TCs. Today, TCs do not form where ocean temperatures are colder than about 26°C. However, temperatures colder than 26°C do not inhibit TC formation. Rather, the 26°C isotherm corresponds to the beginning of the trade wind inversion as one moves poleward, which does directly inhibit the formation of TCs. The equatorward limits of the trade wind inversion are not likely to change significantly as the climate warms, so the area affected by TCs is likely to remain about the same.

The second issue concerns the frequency of TCs. Some AGCM studies indicate a decrease in the frequency of hurricanes as the climate warms, while others indicate an increase (Henderson-Sellers *et al.*, 1998).

The third and fourth issues concern the average intensity of TCs and the intensity of the most severe storms that occur. The most recent simulations, in which a high-resolution hurricane prediction model is embedded in a global AGCM with prescribed SSTs, give an increase in the average strength of a sample of strong hurricanes by 5–12% under climatic conditions associated with a CO$_2$ doubling (Knutson *et al.*, 1998). However, observations of the present climate fail to show any correlation between average hurricane strength and SST, and enough processes are omitted from even high-resolution models that the predictions from such models must still be regarded as very tentative.

Current theories of tropical cyclone development indicate that the maximum potential intensity (MPI) of cyclones increases with increasing SST. The problem is that current theories omit a number of factors, all of which will tend to limit the increase in MPI with increasing SST (Henderson-Sellers *et al.*, 1998). In particular, the cooling effect of increasing evaporation of ocean spray as winds pick up could be a self-limiting process. The question can also be approached empirically. DeMaria and Kaplan (1994) find that the maximum hurricane strength (as opposed to the average hurricane strength) does increase with increasing SST. The problem here is that the temperature of the tropopause also decreases with increasing SST in the SST range where hurricanes occur (from the 26°C isotherm to the equator), and this also tends to increase the maximum hurricane intensity. Unless the correlation between tropopause temperature and SST as climate changes is the same as the spatial correlation observed at present, the observed SST–MPI correlation is inapplicable to a change in climate. Thus, it is not possible to say anything definite about the most destructive hurricanes, based on either theoretical or empirical arguments.

8.7 Linkage between uncertainties in climate sensitivity and changes in mid-latitude soil moisture

In closing this chapter, it is useful to return to the question of the climate sensitivity and the implications of low-latitude climate feedback processes for the impact of global warming on summer soil moisture. Part of the uncertainty over climate sensitivity involves the effect of convection on changes in the amount of water vapour in the upper tropical troposphere. A strong negative feedback due to a downward shift in the distribution of water vapour in this region (due to enhanced subsidence, as suggested by Lindzen, 1990) would reduce the global mean climate sensitivity by making the overall water vapour feedback less positive, but would have its largest impact at low latitudes. As a result, the polar amplification would increase. However, a greater polar amplification of the temperature response results in a greater decrease in mid-latitude summer soil moisture for a given warming in mid-latitudes (Rind, 1987, 1988). Thus, a smaller global mean climate sensitivity does not necessarily mean smaller impacts in the major mid-latitude food-producing regions of the world. The beneficial impacts of a smaller global mean warming could be offset to some extent by more severe impacts on soil moisture. The implications of any substantially revised estimates of climate sensitivity (whether lower or higher) depend on the specific reasons for the revised estimate and the associated changes in the geographical – and especially the latitudinal – pattern of climatic change.

8.8 Reliability of regional forecasts of climatic change

This chapter has emphasized the largest-scale patterns of climatic change expected from increasing GHG concentrations, such as the differences in the zonally averaged response at low, middle, and high latitudes; continental-scale changes in soil moisture; and general tendencies concerning changes in tropical and extratropical cyclones, interannual variability, and monsoon circulations. Many policy analysts and political decision makers, in contrast, are interested in climatic change at the national and subnational scales. However, the smaller the spatial scale of interest, the less reliable are the projections – something that seems to be true in many fields of human enquiry. It is generally agreed that AGCMs and AOGCMs are quite unreliable at the scales of greatest interest to policymakers. Furthermore, it will probably be a very long time before reliable regional forecasts of climatic change can be made. Indeed, it might *never* be possible to make reliable regional-scale forecasts in advance. It is hard to see how the uncertainty in even the global mean forecast of temperature change can be reduced below a factor of 2, given the incredible complexity and multiplicity of interacting feedback processes.

This is not to say that regional forecasts are not without value. Regional forecasts provide an indication of how large the local climatic (and soil moisture) changes can be in association with a given global average warming. That is, they provide an indication of the spatial variability in changes. Even if the forecast change is wrong (in detail) in each and every single region (some regions warming or drying more in reality than forecast, other less), AGCMs and AOGCMs can still provide useful information for the development of GHG emission policies. This is because the range of regional changes seen in a given latitude band for a given model, or the range of changes simulated for a given region by a collection of models, gives an indication of the *risks* associated with climatic change. Policy development on the basis of estimated risks, rather than on the basis of specific regional forecasts (as is often asked for), has a much sounder scientific foundation.

8.9 Summary

This chapter began with an examination of the global mean temperature response to a CO_2 doubling and the related climate sensitivity. The argument that increases in GHG concentrations of the magnitude expected during the 21st century will lead to a significant warming of the climate is based on very fundamental physical principles and observations. Increasing GHG concentrations trap heat, and the heat trapping can be calculated to within ±10% based on laboratory measurements. A trapping of heat must lead to an increase in temperature so as to restore radiative balance. The first benchmark that one can establish is the warming that would be required by a perfect radiator: about 1.0°C for a doubling of CO_2. The second benchmark arises from the observation that the amount of moisture increases throughout the troposphere as the climate warms. This exerts a strong positive feedback, and raises the expected warming for a doubling of CO_2 to at least 1.5°C. This is the second benchmark, albeit less certain than the first benchmark. Cloud feedbacks could further amplify the climate response or could diminish it. Combined with ice and snow feedbacks, this gives an overall range of 2.1°C to 4.8°C in current AGCMs. However, comparisons of paleoclimatic data and of global mean temperature variations during the past 140 years or so both suggest a CO_2 doubling sensitivity within and somewhat below the lower half of the AGCM results.

This chapter also presented the main features of the equilibrium temperature and soil moisture response to a doubling of atmospheric CO_2, as simulated by state-of-the-art AGCMs. Although the quantitative results and regional details differ significantly from one model to another, the large-scale features of the temperature and soil moisture response can be readily understood in terms of basic principles. Changes in other climatic elements of great interest and importance – mid-latitude storms, El Niños, and monsoons – seem to depend critically on the spatial patterns of simulated change. These in turn depend in part on the simulated changes in the ice and snow distribution and on many other details of the models. In addition, the spatial patterns of climatic change during the transition to a warmer climate are likely to differ in important ways from the equilibrium patterns of climatic change. The question of transient or time-dependent climatic change is briefly addressed in the next chapter.

Questions for further thought

1. Explain how the simulated response of soil moisture to a doubled-CO_2 climate is linked to (a) the simulated differences between land and sea warming, (b) the simulated latitudinal variation in the zonally (east–west) averaged warming, and (c) the simulated present-day soil moisture conditions.
2. Which features of the climatic change projected by climate models are expected with (a) high confidence, and (b) low confidence? For each feature, explain why the confidence level is low or high.

Chapter 9

The transient climatic response and the detection of anthropogenic effects on climate

In the preceding chapter we examined the regional patterns of steady-state climatic change following a doubling of atmospheric CO_2, as simulated by AGCMs coupled to simple, mixed layer-only ocean models. A CO_2 doubling was used as a surrogate for the actual combination of GHG concentration changes giving rise to the same global mean forcing as for a CO_2 doubling. Since the focus in these experiments was on steady-state climatic changes, there was no need to consider the delay in surface warming caused by the mixing of heat into the deep ocean. To reduce the computational cost further, the models were subjected to a sudden doubling of the CO_2 concentration once a steady-state climate with the present (or near-present) CO_2 concentration had been reached.

Real climatic change involves the gradual response to gradually increasing concentrations of GHGs. The time-dependent response to a radiative forcing is referred to as the *transient response*. The transient response is relevant to a number of important issues, such as (i) determining the rate of future climatic change and hence the rate at which human societies and natural ecosystems will have to adapt (where possible); (ii) constraining estimates of the climate sensitivity by comparison of simulated global and hemispheric mean temperature changes (ΔTs) during the 20th century with observed temperature trends; and (iii) assessing the extent to which the patterns of temperature and rainfall change computed for a doubling of CO_2 would occur in association with smaller amounts of warming during the transition to a doubled-CO_2 climate.

In order to simulate the transient response to increasing GHG concentrations, the mixing of heat into the deep oceans has to be accounted for. This can be done using the globally averaged, upwelling–diffusion model, or using 2-D and 3-D ocean models. Experiments with the upwelling–diffusion model have provided a number of qualitative insights into the factors controlling the transient response. However, here we focus solely – and briefly – on the horizontal patterns of climatic change as obtained using coupled AOGCMs. We then compare temperature changes during the past 140 years (roughly) as simulated by the 1-D upwelling–diffusion model and as observed. To make a valid comparison requires developing estimates of the impact of natural forcing factors (at the very least, volcanic activity and solar variability) and taking into account the cooling effect of aerosol emissions. In so doing, we develop indirect observational constraints on the magnitude of the aerosol cooling effect and on climate sensitivity.

9.1 Regional patterns of projected transient climatic change

The problem posed by the very long simulation times required to reach a steady state or unchanging climate with coupled AOGCMs has been circumvented in two steps. First, the perturbation experiment with increasing CO_2 does not start from an equilibrium climate, but rather from a spin-up that still has some residual drift. Second, the perturbation simulation is not carried through to a new equilibrium climate either. Rather, the perturbation consists of a 1% per year compounded increase in CO_2 for 70 years or so, and the focus is on the regional patterns of climatic change at the time of a CO_2 doubling (year 70), not the $2 \times CO_2$ equilibrium change. The assumption is that the drift seen in the control climate (owing to the fact that the spin-up simulation was not carried through to equilibrium) can be subtracted from the perturbation simulation to determine the climatic change that would have occurred if there had been no initial drift.

Among the important differences between the steady-state patterns of climatic change (found in experiments with AGCMs) and the transient patterns of climatic change (found in experiments with AOGCMs) are:

- a less than average warming in the North Atlantic and Antarctic Circumpolar Ocean in the transient case, whereas the steady-state case shows a polar amplification (Manabe *et al.*, 1991; Murphy and Mitchell, 1995);
- a greater reduction of mid-latitude soil moisture in summer in the transient case than in the steady-state case because the ratio of land:ocean warming is greater in the transient case, so there is less of an increase in evaporation from the oceans and thus less of an increase in rainfall over land (Manabe *et al.*, 1992);
- significantly different cloud feedbacks between the NH and SH in the transient case due to the greater warming in the NH than in the SH (Murphy and Mitchell, 1995).

Plate 8 shows the latitudinal variation in the zonally averaged surface air temperature response at the time of a CO_2 doubling, as computed by a number of AOGCMs when forced with CO_2 increasing at 1% per year (in most cases, a 10-year average centred around year 70 is given). This figure should be compared with Plate 7 which shows the equilibrium response to a CO_2 doubling. Some of the models show much less polar amplification of the transient temperature response than for the equilibrium response. This can be explained by the greater delay effect of the oceans at high latitudes, where the deepest mixing occurs. Many models show a minimum surface temperature response in the Antarctic Ocean, also due to deep mixing. Figure 9.1 shows the geographical variation of surface-air warming averaged over years 60–80 of the transient simulation of Manabe *et al.* (1991). Also shown is the steady-state warming for a CO_2 doubling as computed using an AGCM–mixed layer model. The lowest panel shows the ratio of transient to steady-state warming, and highlights the limited warming in the North Atlantic and southern circumpolar regions due to deep mixing (indeed, a slight cooling occurs near the Antarctic Peninsula).

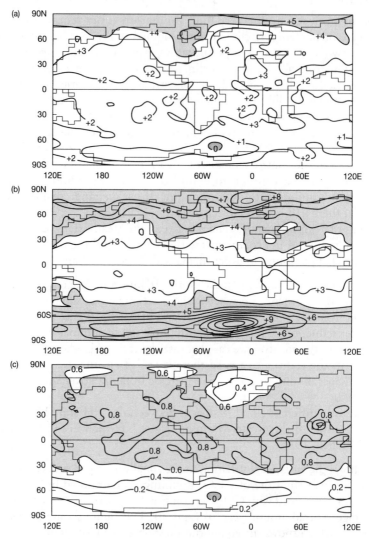

Figure 9.1 (a) The transient surface-air temperature change, averaged over years 60–80 of the AOGCM simulation of Manabe *et al.* (1991), in which atmospheric CO_2 increases by 1%/year starting at year 0. (b) The equilibrium response of surface-air temperature as obtained with the same AGCM, but coupled to a mixed layer-only ocean. (c) The ratio of the transient to equilibrium temperature responses. Reproduced from Manabe *et al.* (1991).

Murphy and Mitchell (1995) found that the differences in the transient surface warming in the SH and NH (averaging 1.1°C and 2.4°C, respectively, by year 70) caused the cloud feedback to be negative in the SH and positive in the NH. This difference in cloud feedback contributed to the hemispheric difference in warming itself. More importantly, the global mean cloud feedback was negative during the early part of the transient response, and positive later on. As a result, the *effective* climate sensitivity was only 1.0°C at year 15 of the simulation, but reached the steady-state sensitivity of 2.8°C by year 40 of the transient response (the effective climate sensitivity is the temperature change that would occur in equilibrium if the initial strengths of all the feedback processes were to persist to equilibrium).

The possibility that the effective climate sensitivity during the transient response to increasing GHGs could be different from the equilibrium climate sensitivity has important implications. This is because efforts have been made to determine the climate sensitivity by comparing model-simulated and observed temperature variations during the 20th century. This approach is valid, however, only if the transient climate sensitivity – which is what is really being constrained – is the same as the equilibrium climate sensitivity. We turn to this issue next.

9.2 Simulating (and therefore explaining) recent past temperature changes

Given an estimate of the climate sensitivity (which, admittedly, can change during the transient), of the uptake of heat by the oceans, and of the radiative forcing due to anthropogenic and natural factors, one can calculate what the variation in global mean temperature should have been during the past century if the forcing factors, climate sensitivity, and oceanic uptake of heat are correct. If the simulated variation closely matches the observed variation, we can say that we have "explained" the observed temperature variation. The degree to which the simulated changes match observations can be compared as the oceanic uptake of heat, uncertain forcings, and the climate sensitivity are varied (all of which can be easily done with simple models). A range of acceptable climate sensitivities can be determined, given that some disagreement between model-simulated and observed climatic change can be expected due to errors in the total forcing, the presence of internal variability in nature, or other factors.

Estimation of the radiative forcing due to GHGs

The radiative forcings due to well-mixed GHGs can be accurately estimated (to within ±15%) based on the known variation in their concentration. On the other hand, the present-day radiative forcing due to changes in stratospheric and tropospheric ozone is quite uncertain (a factor of 2–3 uncertainty), as discussed in Section 6.7. The estimated radiative forcings from pre-industrial times to 1995 were presented in Table 6.2. Figure 9.3(a) gives a middle estimate of the variation in GHG forcing from 1750 to 1990.

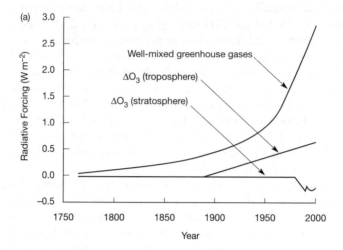

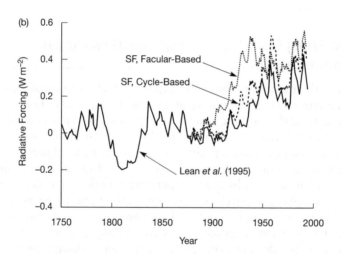

Figure 9.2 Estimates of the globally and annually averaged radiative forcing from 1750 to the present: (a) due to changes in GHG concentrations; (b) due to solar variability, as estimated by Solanki and Fligge (1998) based on either the length of the solar cycle or the number of faculae (bright areas) seen on the sun (SF), and as estimated by Lean *et al.* (1995); (c) due to volcanic activity, based on time series of volcanic activity as reconstructed by Sato *et al.* (1993) and by Robock and Free (1995); and (d) due to anthropogenic aerosols. Two volcanic activity indices by Robock and Free (1995) are shown, one referred to as the Dust Veil Index (DVI), and the other a composite Ice Core–Volcano Index (IVI). For aerosols, three cases are shown, assuming that the total aerosol cooling in 1990 was either 20%, 40%, or 60% of the assumed GHG heating in 1990 of 2.71 W m^{-2}.

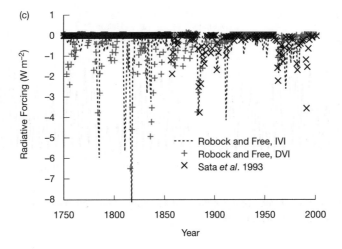

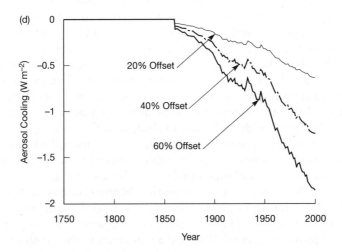

Figure 9.2 Continued

Estimation of the radiative forcing due to solar variability

One of the most striking features of solar activity is the observed variation over time in the number of sunspots. Figure 9.3 shows the observed variation since 1500; an 11-year cycle is evident, as well as longer-term oscillations in the maximum number of sunspots achieved during any given 11-year cycle. However, relating these variations in sunspot activity to changes in solar luminosity (i.e., the energy output from the sun) is uncertain.

There are believed to be two components to the variability in energy output from the sun: a short-term component related to changes in the number of

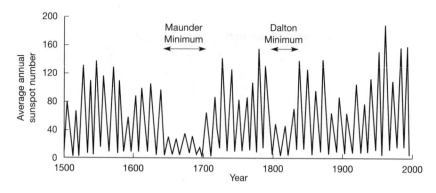

Figure 9.3 Variation in the average number of sunspots visible during each year from 1500 to 1993. Reproduced from Bryant (1997).

sunspots, which has been directly measured by satellites during the past 20 years; and a long-term variation that is derived indirectly from observations of other stars. These two components are referred to as the active sun and quiet sun components, and are denoted by ΔS_{act} and ΔS_{qs}, respectively. The active sun variability is caused by the passage of dark sunspots across the solar disc, which reduce the solar luminosity, and the passage of associated bright areas (faculae), which enhance the solar luminosity. The net result is an increase in solar luminosity with increasing sunspot number, with a luminosity variation of $\pm 1.0 \, \mathrm{W \, m^{-2}}$ during the course of the 11-year sunspot cycle. Long-term variations are thought to be due to changes in the intensity of convective mixing between the deep interior and surface of the sun, and a variety of techniques have been used to estimate how this component has varied over time. Figure 9.2(b) shows the variation of solar radiative forcing[1] as reconstructed by Lean *et al.* (1995), and for two alternative reconstructions by Solanki and Fligge (1998) beginning in 1874. There are some notable differences between the three reconstructions. However, all three time series agree in giving minimum solar luminosity around 1880 and a maximum in the 1930s or 1950s.

A factor complicating the determination of solar radiative forcing is that changes in the ultraviolet flux could induce changes in stratospheric ozone that substantially reduce the variation in net radiation at the tropopause. Haigh (1994) found this to be the case over the 11-year sunspot cycle. The extent to which this would occur in association with longer time-scale solar variability is not known. Nevertheless, it is clear that the correct surface–troposphere radiative forcing due to solar luminosity variations could be substantially smaller than that shown in Figure 9.2(b).

---

[1] The solar radiative forcing, ΔQ, is the change in the global average absorption of solar radiation due to changes in solar luminosity, ΔS. Thus, $\Delta Q = (1.0 - \alpha) \, \Delta S/4$, where α is the average reflectivity of the Earth.

Estimation of the radiative forcing due to volcanic activity

A number of workers have attempted to construct indices of volcanic activity during the past century or longer, based on information from known volcanic eruptions, measurements of atmospheric optical depth, and other indirect information. The climatic effects of volcanic eruptions arise through the injection of SO_x into the stratosphere and the subsequent formation of a hemispheric or global blanket of sulphate aerosols. As the sulphate aerosols are gradually deposited at the Earth's surface, a spike of higher than average acidity appears and is preserved in polar ice caps and glaciers. A number of indices of volcanic activity have been derived from ice cores and other data; these are reviewed by Robock and Free (1995). These indices can be converted to estimates of global mean radiative forcing based on observations made after the 1991 eruption of Mt Pinatubo, which indicate a peak global mean radiative forcing of about $-3.0\,W\,m^{-2}$ about one year after the eruption. Figure 9.2(c) compares the resulting global mean radiative forcing as estimated by Sato et al. (1993) and by Robock and Free (1995). The largest estimated forcing is for the 1815 eruption of Tambora, with a peak annual mean forcing of $-6.5\,Wm^{-2}$ (DVI index) or $-14.9\,W\,m^{-2}$ (IVI index).

Constraining the sulphur aerosol forcing

The forcing due to sulphur + black carbon aerosols is extremely uncertain: after discarding the most extreme estimate, the combined global mean direct + indirect forcing could be anywhere between -0.4 and $-2.6\,W\,m^{-2}$ (Table 6.2). Since the cooling effect of other aerosols could bring the total aerosol forcing to -0.8 to $-3.0\,W\,m^{-2}$, the aerosol forcing would appear to be capable of offsetting most or all of the GHG heating in the global mean. Fortunately, there are two observational constraints that can be used, at least in principle, to limit the magnitude of the aerosol forcing. These are discussed below.

First, during the period 1900–1985, the SH warmed about $0.10 \pm 0.2°C$ more than the NH, in spite of the fact that the lag effect of the oceans should be greater in the SH due to the greater oceanic area in the SH. This difference (if real) can be explained by an aerosol cooling effect, which would be stronger in the NH due to the greater concentration of anthropogenic aerosols. The observed magnitude of the global mean warming and the interhemispheric difference in the warming together set a constraint on the climate sensitivity and on the magnitude of the aerosol cooling effect. If the climate sensitivity is large, then a large aerosol cooling effect is needed to avoid too large a global mean warming, but this in turn increases the difference in the warming between the hemispheres. Calculations performed by Wigley (1989) imply an aerosol forcing in the NH of $-0.7\,W\,m^{-2}$, or about one-sixth of the GHG heating perturbation in the global mean. However, Wigley assumed that aerosols have no cooling effects in the SH. More recent computations of the direct cooling effect of aerosols indicate that the SH mean forcing is 10% (Charlson et al., 1991) to 40% (Schult et al., 1997) as large as the NH mean forcing, and that the indirect forcing, in the SH could be 20% (Boucher and Lohmann, 1995) to 90% (Jones *et al.*, 1994) of that in the NH. When the effect of black carbon aerosols is included, the direct cooling effect is quite small, and there is very little difference in the radiative forcing between the two hemispheres. The primary aerosol cooling effect would then be due to the indirect effect of aerosols on clouds, which tends to have less hemispheric asymmetry than the direct radiative forcing. This in turn would allow a greater aerosol offset of GHG heating than was found by Wigley (1989). Another factor permitting greater aerosol cooling is the fact that the calculated forcing due to tropospheric O_3 is about 40% (Hauglustaine *et al.*, 1994) to 100% (Haywood *et al.*, 1998) larger in the NH than in the SH. However, until the uncertainty in the interhemispheric asymmetry of both the O_3 and aerosol forcing is reduced, it is not possible to say how large an aerosol forcing is permitted under this constraint.

The second potential constraint on the aerosol forcing arises from the observed difference in daytime and nighttime warming over land. As discussed in Section 4.2, average minimum nighttime temperatures over land increased at a rate of 1.79°C per 100 years over the period 1950–1993, while maximum daytime temperatures increased at a rate of only 0.84°C per 100 years. Hansen *et al.* (1995) tested the effect on diurnal temperature variation of a wide variety of heating perturbations, and found that the only mechanism capable of simultaneously generating changes in the mean temperature and in the diurnal amplitude comparable to that observed is localized anthropogenic aerosol and aerosol-induced cloud changes *in combination with* a large-scale warming factor, such as GHG increases. Their analysis implies that the combined aerosol–cloud cooling perturbation is about half as large as the anthropogenic greenhouse heating perturbation in the global mean.

In light of the above discussion, it seems reasonable to consider aerosol forcings that offset 20–60% of the 1990 heating due to GHGs. The temporal variation in the forcing up to (and after) 1990 can then be computed based on the variation in anthropogenic sulphur emissions, the assumed partitioning of the 1990 forcing between direct and indirect effects, and the relationships given in Harvey *et al.* (1997). Figure 9.2(d) shows the resulting variation in the aerosol forcing for cases in which the forcing reaches 20%, 40%, or 60% of the assumed 1990 GHG forcing of 2.71 W m^{-2}, using the sulphur emission estimates given in Figure 3.4. The computed aerosol forcing increases most rapidly after 1950.

The role of vertical mixing in the ocean

When simple climate models such as the 1-D upwelling–diffusion model are used to simulate transient climatic change, the simulated rate of change depends on three parameters or sets of parameters: the radiative forcing (which we have just discussed), the climate sensitivity (which is the key unknown factor that we would like to constrain), and the mixed layer–deep ocean mixing coefficients. The two mixing coefficients in the upwelling–diffusion

Box 9.1 Isopycnal mixing and the effective vertical diffusion coefficient

Mixing in the oceans involves turbulent eddies. Vertical motions are resisted by the density stratification (the layering of less dense water over more dense water), while motions along surfaces of constant density – known as *isopycnal surfaces* – are not resisted at all by the density field. Consequently, the diffusion coefficient – which is the large-scale representation of small-scale turbulence – is several orders of magnitude larger for mixing oriented along isopycnal surfaces than for mixing across isopycnal surfaces (*diapycnal mixing*). The sloping isopycnal surfaces and associated perpendicular line therefore form the natural coordinate system for the representation of diffusion, rather than the vertical and horizontal directions.

Figure 9.1.1 illustrates an isopycnal surface with slope δ. The heat flux along this surface is $k_i \partial T / \partial s$, where k_i is the along-isopycnal diffusion coefficient, and $\partial T / \partial s$ is the temperature gradient on the isopycnal surface. The vertical component of this flux is $\delta / (1 + \delta^2) k_i \partial T / \partial s$. The vertical flux due to diapycnal mixing is equal to $1/(1+ \delta^2) k_n \partial T / \partial n$, where k_n is the diapycnal diffusion coefficient and $\partial T / \partial n$ is the temperature gradient perpendicular to the isopycnal surfaces. Thus, the total vertical heat flux is approximately given by

$$F_z = -\frac{1}{1 + \delta^2} \left[k_n \frac{\partial T}{\partial n} + \delta \frac{\partial T}{\partial s} \right] \tag{9.1.1}$$

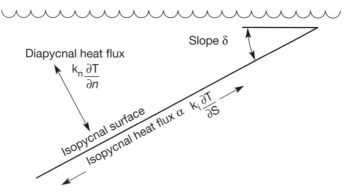

Figure 9.1.1 Illustration of a isopycnal surface, with slop S, and the diapycnal and isopycnal het fluxes.

The effective diffusion coefficient, k_{eff}, is that coefficient which, when multiplied by the vertical temperature gradient, $\partial T / \partial z$, gives the same flux as in Equation (9.1.1). Since $\partial T / \partial z$ and $\partial T / \partial n$ are essentially the same, it follows that

(continued)

(continued)

$$k_{eff} = \frac{1}{1 + \delta^2}\left[k_n + \delta\left(\frac{\partial T/\partial s}{\partial T/\partial n}\right)\right] \qquad (9.1.2)$$

The constant-density surfaces in the ocean tend to be almost exactly parallel to constant-temperature surfaces. Hence, $\partial T/\partial s \approx 0$ and $k_{eff} \approx k_n$ for temperature. For some tracers, such as radioactive carbon and tritium, there are significant gradients on isopycnal surfaces in middle and high-latitude regions because there has been much greater penetration of these tracers into the ocean at high latitudes than at low latitudes. This is illustrated in Figure 9.1.2 for tritium. As a result, $k_{eff} >> k_n$, and the k_{eff} so derived is not applicable to temperature.

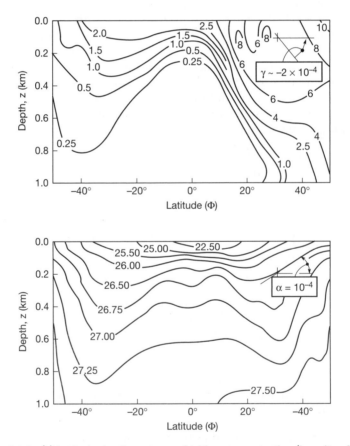

Figure 9.1.2 (a) Latitude-depth contours of tritium concentration (in units of 10^{-15} moles tritium per mole of hydrogen) in the Atlantic basin in 1977 due to bomb testing. (b) Latitude-depth contours of Atlantic basin isopycnals in units of $\Delta\rho = \rho - \rho_o$. (kgm^{-3}) where $\rho_o = 1000$ kgm^{-3}. Reproduced from Hoffert and Flannery (1985).

model are the vertical diffusion coefficient (k) and the upwelling velocity (w). Previous work with the upwelling–diffusion model (and with AOGCMs) has generally used a larger diffusion coefficient than is observed in the real ocean. In the upwelling–diffusion model, k is not a real diffusion coefficient, but rather is an effective diffusion coefficient that represents, among other things, the effects of quasi-horizontal mixing along gently sloping isopycnal (constant-density) surfaces. As explained in Box 9.1, the appropriate diffusion coefficient in the upper 500–1000 m is different for the different properties (tracers) that are being diffused (such as temperature, radioactive carbon, or tritium). More importantly, the effective diffusion coefficient is significantly smaller for temperature than for other tracers. Thus, if k is determined from the mixing of radioactive carbon or tritium (which were both injected into the atmosphere and hence oceanic mixed layer through the atmospheric testing of nuclear bombs) and this k is applied to heat, then the ocean will absorb heat too quickly, causing a greater reduction in the rate of warming. This in turn permits a larger climate sensitivity while still simulating close to the observed temperature increase during the past century or so.

Previous estimates of the permitted climate sensitivity based on the fit of simulated temperature changes to observations have used simple climate models with vertically uniform values of k on the order of $1 \, \text{cm}^2 \, \text{s}^{-1}$. This is much larger, in the upper ocean, than can be justified based on observations. Harvey and Huang (1999) developed an alternative upwelling–diffusion model in which k increases from a value of $0.15 \, \text{cm}^2 \, \text{s}^{-1}$ at the base of the mixed layer to a value of $1.0 \, \text{cm}^2 \, \text{s}^{-1}$ at the 2 km depth. As shown in the next subsection, this results in a smaller permitted climate sensitivity than was obtained before, given the key assumption in all such work that the effective transient climate sensitivity is the same as the equilibrium sensitivity.

Simulated climatic change during the past 140 years

Figure 9.4 shows the variation in global mean temperature from 1850 to 1995 as simulated using the climate model described in Harvey and Huang (1999) and the radiative forcings given in Figure 9.2. The solar forcing of Lean et al. (1995) and the volcanic forcing of Sato et al. (1993) were used for these simulations. The simulations began in 1750, so that the trend by 1850 includes the effect of the forcings during the preceding century, but results are only shown since 1850 because estimates of the variation in global mean temperature begin about then. Results are shown for a steady-state climate response to a CO_2 doubling of 1.0°C, 2.0°C, and 3.0°C. Figure 9.4(a) shows the temperature variation when no aerosol cooling effect is assumed, while Figure 9.4(b) shows the results with offsets in 1990 equal to 20%, 40%, and 60% of the 1990 GHG radiative forcing for climate sensitivities of 1.0°C, 2.0°C, and 3.0°C, respectively. Best results are obtained for a climate sensitivity of 1.0°C with no aerosol forcing. Earlier studies (summarized in Santer et al., 1996; their Section 8.4.1.3) implied that the historical temperature variations are

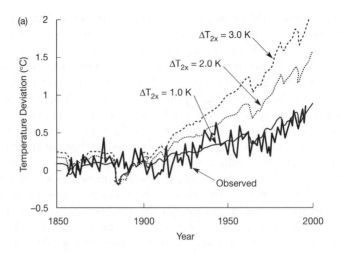

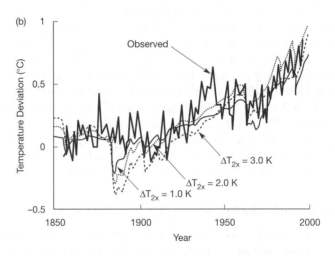

Figure 9.4 Simulated variation in global mean surface air temperature from 1850 to 2000, and comparison with observations (to 1997), for the GHG, solar, and volcanic forcings given in Figure 9.2, and using climate sensitivities of 1.0 K, 2.0 K, and 3.0 K for a CO_2 doubling. Results are given for (a) no aerosol cooling, and (b) aerosol coolings reaching 20%, 40%, and 60% of the 1990 GHG forcing by 1990 for climate sensitivities of 1.0 K, 2.0 K, and 3.0 K, respectively.

consistent with a CO_2 doubling response of 1.5°C and 4.5°C, but, as noted above, the models used in these studies assumed a considerably stronger coupling between the mixed layer and deep ocean than used here.

One very notable discrepancy between the model-simulated temperature variation and the observed temperature variation is the large dip after 1883 in the model results. This is in response to the 1881 eruption of Mt Krakatoa, and subsequent eruptions extending to Mt Katmai in 1912. The peak cooling is modestly greater for larger climate sensitivity, but the greatest effect of higher climate sensitivity is in the long-term response. This behaviour was also found by Lindzen and Giannitsis (1998), using an even simpler model, and suggests a climate sensitivity closer to the lower part of the range considered here.

The following simple back-of-the-argument calculation demonstrates the reasonableness of the 1.0°C climate sensitivity in the absence of aerosol forcing: first, the radiative forcing by 1990 amounts to about $2.6 \, W \, m^{-2}$ due to GHGs plus $0.4 \, W \, m^{-2}$ due to solar variability, for a total of $3.0 \, W \, m^{-2}$; second, this is 80% of the adjusted forcing for a CO_2 doubling of 3.5-$4.0 \, W \, m^{-2}$; and third, the observed warming of 0.65 ± 0.05°C is about 80% of the instant-equilibrium response of 0.8°C that would occur for a climate sensitivity of 1.0°C for a CO_2 doubling. This is a reasonable response.

There are two other considerations that might permit a larger climate sensitivity, apart from the possibility of a partial aerosol offset of the GHG heating. First, there could have been a natural cooling tendency over this period due to an internal oceanic oscillation that masked some of the warming due to increasing GHGs that would have otherwise occurred. Second, the effective climate sensitivity during the transient response, which is really what is being examined here, could be smaller than the equilibrium sensitivity. As discussed in Section 9.1, experiments with the Hadley Centre AOGCM give a much smaller transient sensitivity than the equilibrium sensitivity (1.0°C instead of 2.8°C). Thus, the true equilibrium climate sensitivity could still fall within the upper part of the range considered in Figure 9.4 even without aerosol cooling.

Simulated climatic change during the past 20 years

Hansen *et al.* (1997b) used the GISS AGCM combined with a variety of simple representations of the oceans to simulate climatic variations during the period 1979–1995. They were able to accurately simulate the observed changes in global mean surface, tropospheric, and stratospheric temperatures using forcings due to: (a) the increase in well-mixed greenhouse gases, (b) the decrease in stratospheric O_3, (c) the two climatically important volcanic eruptions (El Chichon and Mt Pinatubo); (d) solar variability; and (e) a constant forcing of $0.65 \, W \, m^{-2}$, which represents the radiative disequilibrium in 1979 due to the fact that the climate at the start of the simulation was not fully adjusted to the preceding buildup of GHGs and aerosols. Plate 9 compares the model and observations (the latter extended here to the end of 1998) for the surface, troposphere, and stratosphere. Among the interesting findings resulting from this work are the following.

- Owing to the cooling effect of the El Chichon and Mt Pinatubo eruptions, there would have been no overall warming between 1979 and 1996 in spite of the continuing increase in GHG concentrations, were it not for the preceding buildup of GHGs and the resulting initial radiative disequilibrium.
- With the initial disequilibrium, the global mean surface warming is 0.1–0.2°C (depending on the ocean model used), which compares well with the observed warming during this period of 0.16°C.

- There is a maximum in the warming in the tropics in the upper troposphere in simulations without loss of stratospheric O_3, but this maximum disappears when O_3 loss is included – in accord with observations.
- The global mean stratospheric cooling between 1979 and 1995 due to the decrease in O_3 alone is 0.6°C, and 0.7°C when changes in well-mixed GHGs are included, both of which are somewhat less than the observed global mean cooling of 0.9°C given by the MSU data (Figure 4.4(a)).

The equilibrium warming for a CO_2 doubling is 3.5°C for the version of the GISS AGCM used here. However, the good fit to observed temperature variations does not validate the high climate sensitivity, since there are relatively small differences in the initial transient response for models with large differences in sensitivity. This is because the relative warming (the absolute warming divided by the final warming) is slower the greater the final warming. Instead, the good agreement with observations serves to validate the processes affecting the vertical structure of temperature changes.

9.3 Summary

This chapter has briefly discussed the time-dependent or transient climatic response to increasing concentrations of GHGs. The salient points to emerge from this discussion are as follows.

- Transient simulations with coupled AOGCMs show substantially different regional patterns of climatic change from those of equilibrium simulations. Since cloud feedbacks depend critically on the changes in atmospheric winds and in vertical motions, which depend on regional patterns of climatic change, the net cloud feedback can be quite different during the transient from that of an equilibrium simulation. In one case where this has been analysed, the effective climate sensitivity during the early transient is only one-third as large as the equilibrium sensitivity.
- Comparison of observed and model-simulated global- and hemispheric-scale temperature changes indicates that much of the warming during the past century is due to anthropogenic effects.
- The observed temperature variations of the last 120 years are best simulated, using the climate model of Harvey and Huang (1999), assuming a climate sensitivity of 1.0°C and no offsetting aerosol forcing. However, given the possibility that the effective transient climate sensitivity could be much smaller than the equilibrium climate sensitivity and that aerosols are offsetting some of the GHG heating, an equilibrium climate sensitivity in line with other lines of evidence (2.0–3.0°C) cannot be ruled out.

Questions for further thought

1. Distinguish between the terms "equilibrium" and "transient" climatic change. Describe the differences in the overall approach that have been used in modelling transient and equilibrium climatic change.

2. What are some of the expected differences between transient and equilibrium climatic change?

3. Distinguish between the terms "equilibrium climate sensitivity" and "effective transient climate sensitivity". What could cause the transient and equilibrium climate sensitivities to differ? Give a specific instance of where this has occurred in climate model simulations.

4. How does the assumed intensity of vertical mixing of heat into the oceans affect the climate sensitivity that is deduced by comparing model-simulated and observed temperature variations during he past century or so?

Chapter 10

Sea level rise

In this chapter the physical processes that determine the rise in sea level that will accompany global warming are discussed, followed by a presentation of recent projections of sea level rise through to the year 2100. Most of the work on possible future sea level rise has focused on the short term (the 21st century). However, the sea level rise issue, perhaps more than any other issue in the climate system response to emissions of GHGs, involves very long time scales and very long lags between the driving factors and the ultimate response of the system. Because of these long lags, an irreversible commitment to substantial sea level rise (several metres) could be made centuries before it occurs. This chapter therefore closes with a discussion of the potential for very large increases in sea level over periods of many centuries to thousands of years.

10.1 Processes tending to cause a rise in sea level

Sea level will tend to increase in association with a warmer climate as a result of (i) thermal expansion of sea water, (ii) melting of small mountain glaciers and ice caps, (iii) melting of the Greenland ice cap, and (iv) changes in the mass balance of the Antarctic ice cap. These processes are amplified below.

Thermal expansion of sea water

As sea water warms, its density decreases. Thus, a given mass of ocean water will occupy a greater volume as the ocean warms, thereby tending to increase the average sea level. The extent of sea level rise depends on the vertical profile of warming within the deep ocean. This in turn depends on the amount of warming in polar regions (which are the source areas for bottom water) and on changes in the strength of the thermohaline circulation. Different ocean models give quite different vertical profiles of temperature change, with significant warming limited to the upper ocean in one case (Mikolajewicz *et al.*, 1990), and greater warming of the deep ocean than at the surface in another case (Manabe and Bryan, 1985). A globally uniform ocean warming by 3°C would cause a sea level rise of about 2.4 m.

Glaciers and small ice caps

Figure 10.1 shows the distribution of mountain glaciers and small ice caps today. Complete melting of all mountain glaciers and small ice caps is estimated to raise sea level by 50 ± 10 cm (Warrick *et al.*, 1996).

Greenland and Antarctica

Greenland is of particular concern because of its large ice mass – sufficient to raise sea level by 7.4 m if it were to completely melt – and the fact that a regional warming of as little as 6°C appears to be sufficient to provoke the eventual melting of the entire ice mass (Letréguilly *et al.*, 1991). Simulations by Crowley and Baum (1995) indicate that, once the Greenland ice sheet disappeared, it would not reform, even if GHG concentrations returned to their pre-industrial values.

Models used to project changes in the mass of both the Greenland and Antarctic ice sheets generally assume that warmer temperatures lead to an increased rate of snowfall. This might not be true, at least in the case of the Greenland ice sheet. Cuffey and Clow (1997) found, based on the characteristics of ice cores from Greenland, that long-term (500–1000 years) accumulation rates have been inversely correlated with temperature during the Holocene period (the last 10,000 years); that is, warmer temperatures have been associated with smaller rates of accumulation. This is explained by shifts in the trajectories of snow-bearing storm systems as climate changes. If this

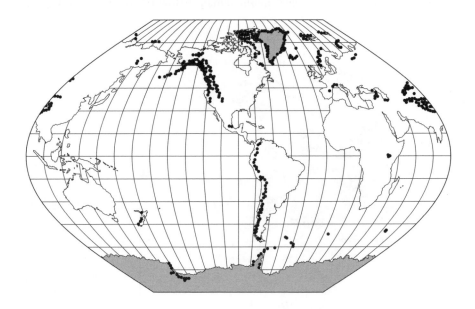

Figure 10.1 Present-day distribution of mountain glaciers and small ice caps, reproduced from Warrick *et al.* (1996).

applies to future changes (and this is a big if), it will exacerbate the near-term sea level rise. The way in which storm tracks change will undoubtedly depend on regional differences in the transient warming, which could be different for rapid, anthropogenically induced climatic change than for the slower, natural climatic changes of the past. The changes in storm tracks could also be different for transient and equilibrium climatic change.

The Antarctic ice sheet can be divided into the East Antarctic Ice Sheet (EAIS), and the West Antarctic Ice Sheet (WAIS), as illustrated in Figure 10.2. Most of the East Antarctic ice sheet sits on bedrock above sea level and, owing to the very cold temperatures, would not be susceptible to summer melting along the edges until the coastal climate had warmed by 6–10°C. At present, the net accumulation of snow on the ice sheet will be largely balanced by flow of ice to the edges and the breaking or *calving* of the ice into the ocean. In contrast, most of the WAIS lies on bedrock below sea level, and large portions of the ice sheet consist of floating ice shelves. The ice sitting below sea level is referred to as *grounded ice*, and the boundary between the grounded ice and the floating ice shelf is referred to as the *grounding line*. The major ice shelves are the Ross Ice Shelf and the Filchner–Ronne Ice Shelf, both of which are fringed by mountainous regions on two sides (see Figure 10.2). The flow of ice from the grounded portion to the ocean is concentrated in fast-moving ice streams. Those flowing into the Ross Ice Shelf are 30–80 km wide, 300–500 km long, and travel at about 0.5 km per year (Oppenheimer, 1998). Mercer (1978) first raised the possibility that melting of the floating ice shelves could cause the grounded ice to collapse into the ocean, raising sea level by 5–6 m. Although the summer air temperature over the ice shelves might remain below the freezing point as the

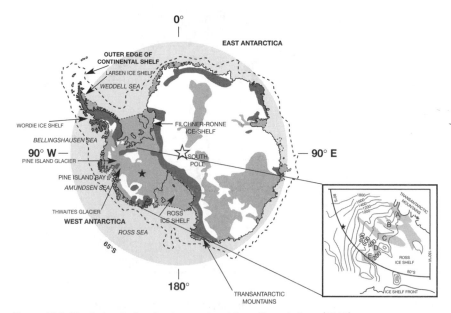

Figure 10.2 The Antarctic ice sheet, reproduced from Oppenheimer (1998).

global climate warms, the ice shelves are thought to be particularly susceptible to melting as the seasonal sea ice around Antarctica retreats and the ocean water adjacent to and under the ice shelves begins to warm.

Oppenheimer (1998) reviewed the latest evidence concerning the stability of the WAIS, and concluded that "the early idea that WAIS is subject to hydrostatic instability and rapid grounding-line retreat seems unlikely, but the issue cannot be conclusively resolved" at present. A key premise underlying concerns about the stability of the WAIS is that regional warming would lead to increased melting at the base of the floating ice shelves. The base of the floating ice shelves occurs at depths as great as 1500 m, and the basal melting that occurs today is driven by the fact that the melting point at the 1500 m depth is –2.9°C rather than –1.9°C, as at the surface. Dense surface water at –1.9°C forms in association with the ejection of a salty brine during the seasonal freezing of sea ice; this water is referred to as "High Salinity Shelf Water" (HSSW). HSSW flows along the underside of the floating ice shelves and, because it ends up warmer than the local melting point, is able to power substantial melting at the base of the ice shelves. Nicholls (1997) presents oceanographic data indicating that a reduction in the rate of formation of sea ice in winter (due to regional warming) would reduce the flux of HSSW, thereby reducing the rate of basal melting. Thus, a warmer climate would, at least initially, lead to an increase in the ice shelf mass, thereby tending to lower sea level. This is in addition to any lowering of sea level that would occur due to an increase in the accumulation of snow on either the EAIS or the WAIS. Warming of the surface water around Antarctica would probably be considerably delayed due to deep mixing of heat into the oceans around Antarctica (Section 9.1).

10.2 Projections of future near-term sea level rise (1990–2100)

The separate contributions of thermal expansion, mountain glaciers and small ice caps, and of Greenland and Antarctica, to changes in sea level have been computed using a variety of climate, glacier, and ice sheet models.

Global mean sea level rise

Table 10.1 summarizes the projected (1990–2100) contributions to global mean sea level rise from the thermal expansion of sea water, melting of mountain glaciers and small ice caps, and from changes in Greenland and Antarctica, as reported in several recently published studies. Also shown is the change in global mean surface air temperature during 1990–2100 associated with each set of results. The various meltwater contributions were computed using either 2-D or 3-D models of the Greenland and Antarctic ice sheets, and a simple book-keeping model of mountain glaciers and small ice caps. For some cases, two or three different temperature changes were considered.

The various approaches yield surprisingly similar results due, no doubt, to the fact that many of the critical underlying assumptions are similar (e.g., no

Table 10.1 Contribution to sea level rise (cm) from 1990 to 2100 due to various processes, as computed by different researchers

Reference	ΔT (°C)	Glaciers and small ice caps	Greenland	Antarctica	Thermal expansion	Total
Wigley and Raper (1995)	2.1	11.9–19.4	—	—	—	—
de Wolde *et al.* (1997)	2.2	12.3	7.2	–7.5	15.2	27
	2.5	14.6	10.4	–8.5	17.5	34
Gregory and Oerlemans (1998)	2.7	13.2	7.6	–7.9	22.5	35
	3.3	18.2	9.3	–9.7	29.9	48
Huybrechts and de Wolde (1999)	1.3	—	3.9	+0.7	—	—
	2.4	—	10.5	–3.0	—	—
	3.5	—	20.8	–6.4	—	—
Bryan (1996)	—	—	—	—	15 ± 5	—

collapse of the WAIS, snow accumulation increasing with temperature). The greatest uncertainty pertains to the response of the Antarctic ice cap. Both de Wolde *et al.* (1997) and Gregory and Oerlemans (1998) find that changes in the Antarctic and Greenland ice masses almost cancel out. This is supported by a study of the effect of a doubled CO_2 climate on the surface mass balance of Greenland and Antarctica by Ohmura *et al.* (1996) using a high-resolution AGCM. In particular, Ohmura *et al.* (1996) find that *changes* in the surface mass balance on the two ice sheets largely cancel out. As noted in Section 4.7, one must assume a substantial contribution (5–14 cm) of Antarctica to sea level rise during the past century in order to explain the observed sea level rise. The results of de Wolde *et al.* (1997) and Gregory and Oerlemans (1998) do not include this possible background trend. In contrast, Huybrechts and de Wolde (1999) simulated the last two glacial cycles (beginning 250,000 years ago) prior to the present, so that the glacier ice mass "inertia" due to past climatic changes is reflected in their projections of future climatic change. They obtained a contribution of Antarctica to sea level rise during the past 100 years of 3.9 cm, which closes one-third to two thirds of the gap between the observed sea level rise of 17–19 cm and the computed rise due to direct anthropogenic effects, the melting of glaciers, and thermal expansion of sea water (5–12 cm). For a constant climate, the projected effect of Antarctica to 2100 is a sea level rise of 5.0 cm. The sea level rises shown in Table 10.1 include this background trend plus the effect of a warming climate, and so the net result is a much smaller tendency for falling sea level than obtained by de Wolde *et al.* (1997) or Gregory and Oerlemans (1998).

The uncertainty in the model-computed background trend is quite large; it could be close to zero, or it could be twice as large. Given the aforementioned discrepancy between observed and computed sea level rise during the past century, the latter seems to be a distinct possibility. For this reason and because of the uncertainty in all of the computed components to sea level rise, it seems reasonable to broaden the original range (27–53 cm by 2100) to 25–70 cm by 2100 if GHG concentrations stabilize at the equivalent of a CO_2 doubling (with a global mean warming of 2.0–3.0°C).

Regional variations in sea level rise

There are marked regional variations in sea level today (by up to 1 m), which give rise to a non-zero slope of the sea surface. Surface ocean currents are in a balance between the Coriolis force that arises from horizontal motion, and the horizontal pressure gradient force created by the sea surface slope. The present-day variation in sea level is related to horizontal variations in the density of sea water (which in turn depend on horizontal variations in the temperature and salinity of sea water), and to horizontal transport of water in the mixed layer due to the applied wind stress (such transport is at right angles to the wind direction, as explained in any introductory oceanography textbook, and is referred to as Ekman transport).

If the ocean experiences horizontal differences in the depth-averaged warming, there will be regional variations in the sea level rise due to the fact that some parts expand more than others, and the water in the regions that expands more does not, by and large, flow outward to the other regions. Rather, an adjustment in the surface ocean currents will occur so as to largely maintain the initial variations in the sea level rise. Regional variations in sea level rise will also occur as a result of changes in the surface wind strength, as this will alter the Ekman transport of surface water.

Bryan (1996) computed the variation in sea level rise for a transient run of the GFDL AOGCM in which the atmospheric CO_2 concentration increases by 1% per year. The greatest warming occurs in the North Atlantic Ocean, owing to the deep mixing of surface heat (this deep mixing also causes the minimum in surface-air warming in this region, seen in Figure 9.1). Figure 10.3 shows the regional variation in sea level rise averaged over years 70–79; only the contributions due to thermal expansion of sea water and changes in ocean currents are represented in this figure. The average sea level rise is 15 cm, but differs by in excess of 15 cm from one region to another. The greatest sea level rise occurs in the North Atlantic Ocean and along the east coast of North America, owing to the markedly greater warming of the ocean in this region. However, the greater warming at depth in the North Atlantic Ocean than at lower latitudes reduces the horizontal density contrast, which drives the thermohaline overturning. The thermohaline overturning and the associated poleward transport of heat thus decrease in strength. This in turn reduces the ocean warming at high latitudes and increases it at low latitudes, thereby reducing the difference in sea level rise. Thus, the main smoothing out of the horizontal variations in sea level rise occurs through a smoothing out of the temperature changes due to induced changes in ocean currents, rather than through a redistribution of ocean mass.

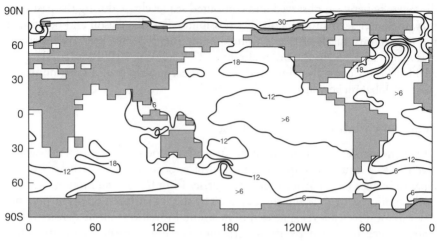

Figure 10.3 Regional variation in sea level rise (cm), averaged over years 70–79 of an experiment by Bryan (1996), in which CO_2 increases by 1% per year compounded. Only the effects of thermal expansion of seawater and changes in ocean currents are shown here.

An interesting consequence of the non-uniform sea level rise is that the sea level rise averaged over all coastlines can be quite different from the global mean sea level rise (Hsieh and Bryan, 1996). This complicates the comparison of tidal gauge records of sea level rise with model-predicted sea level rise.

10.3 Prospects for long–term sea level rise

As noted in the introduction to this chapter, the sea level response to climatic change involves very long time scales and entails commitments that will endure for centuries after consumption of fossil fuels has ended. The sea level rise projected for the end of the 21st century – 25 to 70 cm – is likely to be only a small fraction of the sea level rise that will ultimately occur, even if GHG concentrations are rapidly stabilized during the early part of the 21st century.

Thermal expansion of sea water

Experiments with a 2-D ocean model by Harvey (1994) indicate that a 3.0°C global mean surface warming could be associated with an eventual sea level rise of less than 0.5 m to almost 3.0 m, depending on how the temperature change varies with depth. The time to reach 50% of the final sea level rise ranged from about 200 to 1000 years, while the time required to reach 75% of the final sea level rise ranged from about 400 to 2400 years.

Greenland and Antarctica

Complete melting of Greenland, which could be triggered by as little as an equivalent CO_2 doubling, would lead to an increase in sea level by about 7 m, over a period of 1000–5000 years. Collapse of the West Antarctic Ice Sheet, which could also be triggered by as little as an equivalent CO_2 doubling, would lead to a sea level rise of 5–6 m, over a time period of 1000–2000 years. The melting of East Antarctica, on the other hand, would require a regional warming on the order of perhaps 20°C, which is not likely except under scenarios of very high GHG emissions combined with moderate to high climate sensitivity.

Total sea level rise

Based on the above discussion, it can be seen that a sustained increase in GHG concentrations equivalent to a CO_2 doubling runs a significant risk of provoking an eventual sea level rise of up to 14–17 m. The component of this sea level rise due to melting of ice would be reduced by up to one-third by the tendency for the oceanic crust to sink under the weight of the additional water mass. This would occur on a time scale of several thousand years, which is comparable to the time required for large changes in the Greenland and Antarctic ice masses. Peak sea level rise may therefore be on the order of 10–13 m, which is still substantial.

10.4 Summary

Based on the evidence presented in this chapter, it is concluded that (i) global mean sea level is likely to rise by 25–70 cm by the year 2100, and (ii) a sustained increase in GHG concentrations equivalent to a doubling of atmospheric CO_2 runs a strong risk of an increase in sea level of 10–13 m over a period of several thousand years.

Questions for further thought

1. Summarize the factors contributing to projected future sea level rise in terms of (a) magnitude, (b) time scale, and (c) degree of uncertainty.

Part III

The science–policy interface

The first two parts of this book have brought together information from the disparate disciplines that are needed to understand how the climate system responds to anthropogenic emissions of GHGs and aerosols and their precursors. We have examined the emission factors that link human activities to gas emissions, and have traced the sequence of events and processes involved in the carbon cycle and other biogeochemical cycles. These cycles determine the buildup in concentrations, from which one can calculate the radiative forcing. We then examined the transient and steady-state climate response to the projected radiative forcing, with an emphasis on those results that are independent of the details of any given simulation model. Lastly, we considered the effect on sea level rise. We have thus built an understanding of the processes involved, in nature, in translating human emissions of gases into physical impacts – changes in climate and changes in sea level at the regional level.

The development of a policy response to the prospect of global warming and rising sea level requires much more than the consideration of physical impacts. The anticipated changes in climate and sea level need to be translated into a wide array of social, economic, and biological impacts – changes in food production, in forests, in human welfare and well-being, in the health and vitality of natural ecosystems. Many of the impacts have costs that cannot be quantified in economic terms. The valuation of many impacts – particularly those involving loss of species and ecosystems – will differ from individual to individual, among different cultures, and probably from one generation to the next. Most of the economically quantifiable costs depend on the extent and ease of adaptation, which in turn depends on the nature of society and technology by the time the changes occur. There is a wide divergence of opinion concerning the costs of climatic change, as reflected in reviews and discussions by Pearce *et al.* (1996) and Demeritt and Rothman (1998a,b).

The development of a policy response also requires consideration of the costs (and non-climatic benefits) of actions to reduce emissions of GHGs. In the case of energy-related emissions of GHGs, this entails improving the efficiency with which energy is used and transformed beyond the efficiency improvements that would occur anyway, increasing the provision of non-fossil fuel sources of energy beyond that which is expected to occur anyway, possibly reinjecting some CO_2 into the ground or deep into the ocean, and implementing broad-based policies to stabilize the human population at the lowest possible level. Many of these options and the associated cost estimates are reviewed in Ishitani *et al.* (1996), Kashiwagi *et al.* (1996), Michaelis *et al.* (1996), and Levine *et al.* (1996). Options to limit population growth, and related issues, are discussed by Bongaarts (1990, 1994). The reduction of land-use related emissions of GHGs entails developing sustainable forestry and agricultural practices (Brown *et al.*, 1996). In the case of agricultural emissions of methane and N_2O, this entails modifying current practices in ways that amount to increasing the efficiency with which fertilizers and animal feed are used (Mosier *et al.*, 1998a,b). As in the projected costs of climatic change, there is an enormous range of opinion concerning what the costs of these actions will be, with the anticipated costs being highly dependent on somewhat arbitrary assumptions concerning the ease with which new technologies can be introduced, the future cost of non-fossil fuel energy sources, and a whole array of social and economic factors (Repetto and Austin, 1997). Climatic change and efforts to limit emissions both entail risks (Harvey, 1996a,b). The development of a policy response to climatic change requires placing all the available information in a risk-minimizing framework.

Finally, the development of a policy response requires consideration of the politics of climatic change – the fact that an effective response requires coordinated, broadly based international action involving countries with widely differing economic and social structures and widely differing interests. One must take into account the *process of negotiating* effective international agreements to limit emissions of GHGs, the *structure* of such an agreement once in place, and the range of international *policy instruments* available to ensure effective and equitable international participation in any agreements to limit emissions of GHGs (Sebenius, 1991; Grubb, 1992).

It is far beyond the scope of this book to address any of the three issue areas identified above. Rather, in this concluding part, we shall use the tools developed in Parts I and II to present current scientific understanding in a way that is useful for the development of limits on GHG emissions at the global scale and in the development of strategies to deal with the threat posed by unrestricted emissions of GHGs. We shall do this by using simple models to generate scenarios of climatic change for a range of possible future emissions of GHGs and aerosol precursors.

Chapter 11

Scenarios of future climatic change

In order to explore the interface between science and policy, a series of policy-oriented questions will be addressed in this chapter. The purpose here is not to provide answers to specific policy questions, but rather to show how the scientific information presented in this book can be properly used as an *input* to the process of developing a policy response to the prospect of large, anthropogenically induced climatic change. Prior to doing this, we must first address the question "What is the appropriate scale of analysis?" That is, can useful policy insights be gained through the analysis of global-scale emissions and global-scale climatic changes, or must the policy analysis be grounded in changes at the regional level? Or is some mixture of the two scales of analysis most appropriate?

11.1 What are the appropriate scales of analysis?

Anticipated climate change is expected to vary greatly from region to region, and from season to season. First, the warming of the climate is expected to be largest at high latitudes and smallest at low latitudes. At high latitudes, the warming is expected to be greatest in winter (due to the thinning of sea ice) and smallest in summer. Elsewhere, the seasonal variation is expected to be much smaller, with greatest warming occurring in spring in some regions (due to earlier retreat of snowcover) and in summer in some other regions (due to reduced soil moisture). Within the broad latitudinal zones, however, there is expected to be much variability in the temperature and moisture response. The former was illustrated in Figure 8.2, which shows large variations in the projected equilibrium summer or winter temperature change for a CO_2 doubling at any given latitude.

The impacts of climatic change in any given region depend on the specific climatic changes that occur in that region. As noted above, local climatic changes can differ substantially from the globally averaged climatic change. The net global impact of climatic change will be given by the summation of regional impacts (modified by international trade, in the case of economic impacts), driven by regionally heterogeneous changes in climate. A consideration of climatic change at the regional level would therefore appear to be mandatory in the development of both national-scale and global-scale policy responses to the prospect of global warming. Indeed, this imperative has been driving the effort to develop better predictions of climatic change at the regional level.

There is a fundamental problem: the information that is most needed in order to assess the impacts of climatic change (namely, change at the regional level) is also the least reliable output from computer simulation models. This is evident from Figure 8.2, where large differences between the different models can be seen in many regions. The predicted temperature change in a given region differs by up to a factor of 6. This is much larger than the difference in the predicted global mean temperature response among the models shown in Figure 8.2, which ranges from 2.5°C to 4.0°C – less than a factor of 2 (among all the models shown in Table 8.1, the range in global mean warming is somewhat more than a factor of 2). It is a basic property of the climate system that the uncertainty in projected change is larger at the regional scale than at the global scale. This is because some of the differences between different models at the regional scale cancel out when summed globally. Since it is difficult to see how the uncertainty in the global mean temperature response can ever be reduced below a factor of 2 (at best), it is also unlikely that the uncertainty in the predicted climatic change will ever be reduced much below that seen in Figure 8.2. This is especially the case for transient (time-evolving) climatic change, which introduces another source of uncertainty beyond that applicable to the equilibrium temperature changes shown in Figure 8.2: the impact of regionally varying rates of uptake of heat by the oceans, and the repercussions of this for changes in ocean currents, atmospheric wind patterns, and regional cloud feedbacks. Changes in all of these could be quite different from the equilibrium changes, so several additional degrees of freedom (and associated uncertainty) are added to the transient response.

In light of these uncertainties, what is the value of assessments of the impact of climatic change at the regional level? Regional-scale projections of climatic change, and associated impact assessments, indicate the *risks* that climatic change poses for specific regions. They indicate the potential *magnitude* of local climatic change that could be associated with the generally much smaller, global mean climatic change. However, there is no scientific justification for developing policy responses at the national level based on the climatic changes and impacts within the jurisdiction under question, as given by any one climate model. Rather, the scientific understanding of anthropogenic climatic change can be used to support a *collective* (i.e., global-scale) policy response based on *generalized risks*, rather than nation-by-nation policy responses based on expected nation-by-nation impacts. In any case, the latter approach would not be effective, since emission reductions by one nation are of little value in reducing the buildup of GHGs unless all or most nations undertake similar actions.

This thinking is implicit in the wording of the United Nations Framework Convention on Climate Change (UNFCCC) and in the Kyoto Protocol, that was agreed to in December 1997. Nowhere do either of these agreements contemplate preventative policy responses based an anticipated impacts within the jurisdiction contemplating preventative responses. Rather, the UNFCCC refers to "common but differentiated responsibilities and respective capabilities" (Article 3.1).

In short, the role of regional-scale analysis is to indicate the level of risk, and the potential magnitude of local climatic change, associated with a given global mean warming. The global mean warming can therefore be thought of as an *index* of the risks posed by a particular scenario of GHG emissions, since regional impacts can be expected, on average, to increase with increasing global mean temperature.[1] This provides the justification for focusing on global-mean climatic change, as projected by globally averaged models, in assessing the merits of alternative scenarios of global-scale emissions.

11.2 Simulation methodologies

Two different methodologies are possible in analysing emissions and climatic changes at the global scale. First, the consequences for climate and sea level of a set of scenarios of future emissions of GHGs and aerosol precursors can be computed through a series of *forward calculations* – proceeding from emissions, to concentrations, to radiative forcing, to climatic change and sea level rise, in the same sequence as presented in Parts I and II of this book. Second, a set of upper limits for the concentration of CO_2 and other GHGs can be adopted, involving eventual stabilization of concentrations, and the time-varying permitted emissions can be worked out using an *inverse calculation*. The mathematical procedure involves repeated calculations on each time step until no further change in the computed emission for that time step occurs, a procedure that would be prohibitively time-consuming if high-resolution models were used. The associated climatic change and sea level rise can be deduced using the usual forward calculations, given the prescribed variation in concentrations. Given a set of assumptions concerning the growth of the human population and of average per capita income (the product of which gives the growth of the global economy), the required rates of improvement in the overall efficiency with which energy is used and in the rate of introduction of non-fossil energy sources can be computed, and the tradeoffs between these two options examined.

Using the forward approach, the following issues can be addressed: (i) the overall consequences of different scenarios of fossil fuel CO_2 emissions, (ii) the relative climatic benefits, and the timing of these benefits, when greater efforts are made to reduce emissions of different GHGs, and (iii) the tradeoffs involved when simultaneous reductions of emissions of GHGs and aerosol precursors occur.

---

[1] The global mean warming seems to be a good index of overall risks when the warming is increasing due to increasing forcing, with a roughly constant spatial pattern of warming. However, as discussed in Section 8.7, lower global mean warming due to a downward revision in the climate sensitivity may not imply reduced impacts in mid-latitude regions if the lower climate sensitivity is due to reduced warming in the tropics, since there would be a smaller increase in the moisture supply to mid-latitudes in this case.

The usefulness of the inverse approach stems from the fact that the UNFCCC declares its ultimate objective to be "stabilization of greenhouse gas concentrations in the atmosphere at a level that would prevent dangerous anthropogenic interference with the climate system". The UNFCCC does not specify what concentrations constitute "dangerous" interference, but the Convention does establish three broad objectives which are to be satisfied by whatever stabilization ceilings are adopted: (i) stabilization is to be achieved, "within a time frame sufficient to allow ecosystems to adapt naturally to climate change"; (ii) it should "ensure that food production is not threatened"; and (iii) it should "enable economic development to proceed in a sustainable manner" (Article 2).

Simple climate models such as the 1-D upwelling–diffusion model can be combined with a globally aggregated model of the terrestrial biosphere, and used for both forward and inverse calculations. Such models are ideal tools for global-scale policy analysis, because the key parameters that govern the model response – such as the climate sensitivity, the intensity of ocean mixing, and the CO_2 fertilization effect – can be easily altered to reflect a wide range of possibilities concerning the behaviour of the real world. For each set of physical and biophysical parameters, a wide range of emission scenarios can be used. However, the number of cases that could be considered quickly multiplies, and simple models provide the only practical means for examining the cases of greatest interest. The use of simple models, however, restricts the analysis to the global scale, because such models are either globally averaged or have very coarse spatial resolution. However, we have already argued (in Section 11.1) that this is the most appropriate scale for the analysis of alternative emission scenarios.

11.3 Forward calculations using simple models

In this section, examples of the forward calculation approach are presented. This is followed in the next section by examples of inverse calculations.

Scenarios of future GHG and aerosol emissions

The projection of future GHG and aerosol emissions is fraught with enormous uncertainty, inasmuch as future emissions depend on a wide range of economic, sociological, and technical conditions. Not only are some of these conditions inherently unpredictable, but many of them – such as future levels of energy efficiency, lifestyles, and the form of new urban areas – are subject to deliberate choice. Projections of future GHG and aerosol emissions should therefore not be regarded as predictions, but rather as scenarios showing what the future would be like for specific sets of assumptions.

Figure 11.1 shows projections of fossil fuel emissions of CO_2 and total anthropogenic emissions of CH_4, N_2O, and SO_2 for a typical business-as-usual (BAU) scenario, a GHG emission stabilization scenario, and a GHG emission reduction scenario. Adoption of these scenarios does not imply that any one of them is necessarily feasible or desirable. The BAU CO_2 scenario used here is the IPCC IS92a scenario (Leggett *et al.*, 1992), in which global fossil fuel CO_2 emissions increase from about 6.0 Gt C per year in the 1990s to 20 Gt C per year by 2100. Methane emissions increase from 263 Tg CH_4 per year to 524 Tg CH_4 per year, while N_2O emissions increase from 6.25 Tg N per year to 12.6 Tg N per year.

Future sulphur emissions will depend on the degree of concern over acid rain, since sulphur emissions are the single most important precursor to acid rain. If, as many economists project, the entire world becomes richer during the 21st century, then sulphur emissions can be expected to decrease even if CO_2 emissions rise dramatically, since wealthier people tend to demand – and can afford – a cleaner environment. As noted in Section 3.1, the technology

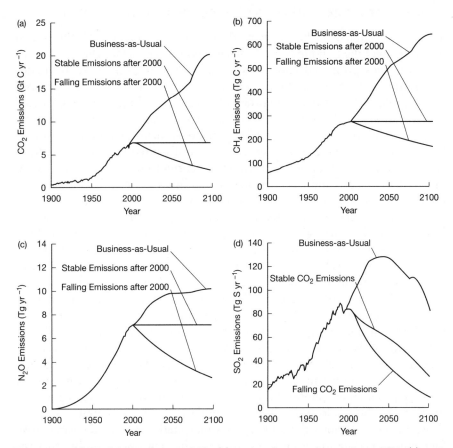

Figure 11.1 (a) Fossil fuel emissions of CO_2, (b) total anthropogenic emissions of CH_4, (c) total anthropogenic emissions of N_2O, and (d) anthropogenic emissions of SO_2, for a typical business-as-usual scenario, a scenario in which GHG emissions are stabilized, and an aggressive GHG emission reduction scenario.

already exists to decouple S and C emissions. Given that programmes are already in place to significantly reduce S emission in North America and Europe, and that current S emissions in Asia are already causing serious acid rain problems (Mohan and Kumar, 1998; Qian and Zhang, 1998), it is unlikely that future total global emissions will rise much above present levels, and they could very well decrease substantially. For all the SO_2 emission scenarios shown in Figure 11.1, it is assumed that the S:C emission ratio decreases linearly to 25% of its year 2000 value by 2100. For the base case CO_2 emission scenario, absolute S emissions increase by about 50% before returning to the current level by 2100. If CO_2 emissions are reduced relative to the BAU scenario, then absolute S emissions fall more rapidly, but this is consistent with the greater overall concern over environmental issues that is implied by a CO_2 emission reduction scenario.

Scenarios of future GHG concentrations and radiative forcing

The coupled climate–carbon cycle model of Harvey and Huang (1999) is used here to project the buildup in the concentration of GHGs for the set of emission scenarios presented above. As noted in Section 7.3, there are major uncertainties concerning the present role of the terrestrial biosphere as a carbon sink, such that the current rate of uptake of anthropogenic CO_2 could lie between 0.5 and 4.5 Gt C yr^{-1}. As discussed in Section 7.8, the uncertainty in the current oceanic rate of uptake appears to be smaller, with most estimates lying between 1.4 and 2.6 Gt C yr^{-1}. Finally, as discussed in Section 3.2, there are substantial uncertainties concerning the current rate of emission of CO_2 due to deforestation and land degradation, and how these emissions varied in the past. In order to simulate the observed buildup of atmospheric CO_2 up to the present, a larger terrestrial biosphere sink is required the larger the assumed land use emission, for a given oceanic uptake. Thus, land use emissions of 0.5 and 1.5 Gt C yr^{-1} by 1990 require a terrestrial biosphere sink of 0.7 and 1.7 Gt C in 1990, respectively, given an oceanic sink in 1990 of 2.0 Gt C. The size of the terrestrial biosphere sink in turn can be altered by changing the value of the model parameters that govern the response of photosynthesis and respiration to increasing atmospheric CO_2 and temperature, within the range of observational uncertainty. Whatever the current CO_2 emission rate due to deforestation, the rate will decrease substantially within 20–30 years, if for no other reason than that there will be fewer forests left. However, since a greater rate of deforestation emissions today implies a greater responsiveness of net primary production (photosynthesis minus respiration) to increasing atmospheric CO_2, this results in smaller future projected CO_2 concentrations.

Figure 11.2(a) shows the buildup of atmospheric CO_2 for the three fossil fuel emission scenarios given in Figure 11.1, in each case assuming that the deforestation emissions were 0.75 Gt C yr^{-1} in 1990 and decrease to zero by 2020. Figures 11.2(b) and 11.2(c) show the buildup of methane and N_2O, respectively, using the corresponding emission scenarios given in Figure 11.1.

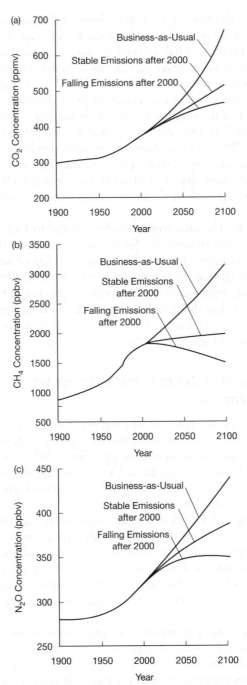

Figure 11.2 Variation in atmospheric concentration of (a) CO_2, (b) CH_4, and (c) N_2O for 1950–2100 for the corresponding emission scenarios in Figure 11.1. CO_2 concentrations were computed using the carbon cycle model of Harvey and Huang (1999), while CH_4 and N_2O concentrations were computed using the assumptions given in Harvey *et al.* (1997).

With regard to the forcing by halocarbons and changes in stratospheric ozone, emission scenarios similar to that used by Solomon and Daniel (1996) are adopted. The buildup of halocarbons is computed using Equation (7.3) (F and τ now pertaining to whatever gas is being considered) and the atmospheric lifetimes and radiative forcings as given in Table 7.1 of Harvey (1999). The effective stratospheric chlorine concentration peaks around year 2000, as does the negative radiative forcing due to loss of stratospheric ozone. As shown by Solomon and Daniel (1996), the forcing due to halocarbons peaks at around $0.3\,\mathrm{W\,m^{-2}}$ in year 2000, but declines much more slowly (as CFC concentrations fall) than the recovery in stratospheric ozone. This is due to the fact that the HFCs have no ozone-depleting effect but are strong GHGs, and the heating effect due to the growth in their concentration largely compensates for the decrease in CFCs.

Figure 11.3 shows the variation in the forcing due to CO_2 alone, due to all other GHGs, and due to aerosols, for the concentration scenarios given in Figure 11.2. The radiative forcing due to the buildup of aerosols was computed by assuming (i) that the forcing in 1990 was 20%, 40%, or 60% of the 1990 GHG forcing, with one-third due to direct effects and two-thirds due to indirect effects, (ii) that the direct forcing varies in direct proportion to the total S emission (given in Figure 11.1(d)), and (iii) that the indirect forcing varies with the natural logarithm of the total emission.

Simulation of future rates of climatic change for surprise–free scenarios

The next step is to use the radiative forcings given in Figure 11.3 to drive a climate model that takes into account the absorption of heat by the oceans. Here, the same quasi-one-dimensional upwelling–diffusion model that was used to compute the oceanic uptake of CO_2 is used for this purpose. Since all models, particularly the 1-D upwelling–diffusion model, are limited in their ability to simulate abrupt climatic changes or "surprises", the results obtained with the 1-D model will be referred to as "surprise-free" scenarios. Reality may very well include a number of surprises, either pleasant or unpleasant. The likelihood of surprises is likely to increase the greater and the faster warming occurs, and this should be kept in mind when viewing the results presented here. The prospects for surprises in the climate system response to increasing GHG concentrations, and their policy implications, are briefly discussed at the end of this chapter.

The major uncertain parameter affecting future climatic changes, for a given emission scenario, is the climate sensitivity. Whatever the assumed climate sensitivity, the model should be constrained to roughly simulate the observed increase in temperature over the past 140 years. This in turn requires assuming a larger present-day aerosol cooling effect the larger the assumed climate sensitivity (otherwise, the larger the assumed climate sensitivity, the greater the simulated temperature change). As shown in Section 9.2, a reasonable simulation of the observed variation in global mean temperature during the past 140 years is obtained if the three aerosol forcing cases given above (whereby the forcing reaches 20%, 40%, or 60% of the GHG heating by 1990) are used in combination with climate sensitivities for a CO_2 doubling of 1.0°C, 2.0°C, and 3.0°C, respectively. Carbon dioxide doubling sensitivities of 1.0°C and 3.0°C represent reasonable upper and lower limits to the climate response, while 20% and 60% aerosol offsets (in the global mean) represent reasonable upper and lower limits to the aerosol forcings. The base-case ocean

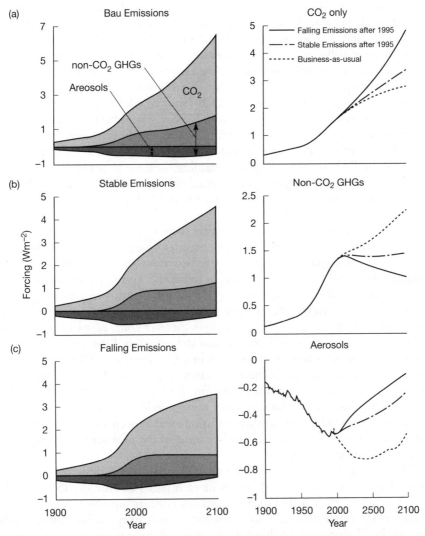

Figure 11.3 Global mean radiative forcing due to (a) CO_2 alone, (b) all GHGs except CO_2, and (c) sulphate aerosols, for the concentration scenarios given in Figure 11.2. In the case of aerosols, the forcing was computed as explained in the text.

mixing parameter values from Harvey and Huang (1999) are used, and the thermohaline overturning is assumed to decrease by 10% per degree Celsius of mixed layer warming. This has negligible effect on the CO_2 buildup but slightly slows down the transient temperature response compared to the case with fixed overturning intensity.

Figure 11.4 shows the projected change in global mean surface air temperature for the three climate sensitivities and for the three emission scenarios. For the case of low climate sensitivity (Figure 11.4(a)), the impact of imposing constant or falling emissions after year 2000 is evident by 2010. For the case of high climate sensitivity (Figure 11.4(c)), the negative aerosol forcing is assumed to be three times larger than for the case with 1.0°C climate sensitivity, and the reduction in the aerosol cooling effect when CO_2 (and other GHG) emissions are constrained initially more than compensates for the reduced GHG heating. This is because the decrease in aerosol cooling occurs immediately as emissions are reduced, owing to the very short lifespan of aerosols in the atmosphere, whereas the impact of reduced emissions of CO_2 (and, to a lesser extent, most other GHGs) does not appear until after a few decades, when there is finally a noticeable difference in the cumulative buildup. As result, the transient warming for the CO_2 emission stabilization or reduction scenarios is initially slightly *larger* than for the BAU scenario. Reduced warming for these scenarios is not evident until 2040. These results are similar to those of Wigley (1991), who was the first to show that simultaneous controls on CO_2 and SO_x emissions could initially cause greater climatic warming.

The above results appear to imply that there is little climatic benefit from reducing GHG emissions before several decades have passed, because there is a coupling in the scenarios presented above between CO_2 emissions and aerosol cooling (reduced CO_2 emissions lead to reduced S emissions and an immediate reduction in the cooling tendency). However, one can recast the apparent tradeoff in an entirely different light by postulating a series of increasingly stringent reductions in sulphur emissions, *independently* of what happens to GHG emissions. One can then examine the extent and the time frame over which increasingly stringent reductions in CO_2 alone, non-CO_2 GHGs, and all GHGs together can offset the *extra* warming tendency due to the reduction in sulphur emissions. The value in isolating CO_2 and non-CO_2 GHGs is that many of the latter have relatively short atmospheric lifetimes, so that reduced warming due to reductions in their emissions will be seen sooner, thereby providing an opportunity to partially offset some of the near-term extra warming due to reduced sulphur emissions.

Figure 11.5 shows the results of a series of experiments designed to answer this question. The three solid curves in each panel assume the IS92a GHG emissions, but with different assumptions concerning sulphur emissions. Under BAU conditions, the S:C emission ratio is likely to fall substantially during the 21st century. To illustrate the effect of this anticipated reduction on global mean temperature change, a constant S:C emission ratio is adopted as the base case. The other two solid curves show the temperature change when the S:C ratio drops to 25% or 6.25% of the 1990 value by 2100; the former

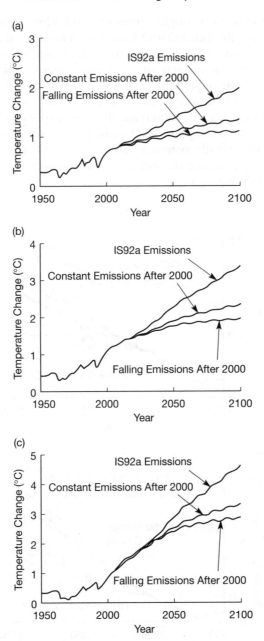

Figure 11.4 Scenarios of future climatic change, corresponding to the radiative forcing variations given in Figure 11.3, which in turn correspond to the emission scenarios given in Figure
11.1. Results are given for (a) a climate sensitivity of 1.0°C for a CO_2 doubling and an aerosol
cooling equal to 20% of the GHG heating in 1990, (b) a climate sensitivity of 2.0°C for a CO_2
doubling and an aerosol cooling equal to 40% of the GHG heating in 1990, and (c) a climate
sensitivity of 3.0°C for a CO_2 doubling and an aerosol cooling equal to 60% of the GHG heating in 1990. In all cases the S:C emission ratio decreases to 25% of its 1990 value by 2100.

holds global S emissions roughly constant, and allows substantially more warming than for the (unrealistic) base case. The three dashed curves all assume the intermediate S emission scenario (in which S:C drops to 25% of the 1990 value), but increasingly stringent restrictions on GHG emissions or ozone precursors. The uppermost dashed curve shows the effect of assuming that the emission ratio of ozone precursors to CO_2 decreases to 6.25% of the 1990 level (i.e., that there is a substantial effort to address problems of ground-level ozone) and that CH_4 and N_2O emissions are stabilized. The middle dashed curves additionally assume stabilization of CO_2 emissions, while the lower dashed curves assume the scenarios of declining CO_2, CH_4, and N_2O emissions shown in Figure 11.1.

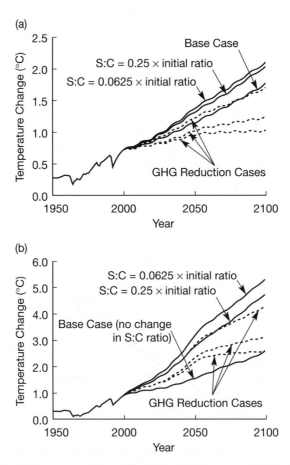

Figure 11.5 Global mean temperature change for business-as-usual GHG emissions and a constant S:C emission ratio, or a ratio decreasing to 25% or 6.25% of the 1990 value by 2100 (solid lines); and for the same absolute S emissions as for the 25% case, but with increasingly stringent controls on GHG emissions (dashed lines). Results are given for a global mean equilibrium warming for a CO_2 doubling of (a) 1.0°C, and (b) 3.0°C.

The results shown in Figure 11.5 show that, if aerosols are currently offsetting only 20% of the GHG heating effect, then reductions in non-CO_2 GHGs alone can offset the extra heating effect (in the global mean) due to stabilization of global S emissions. For the case in which aerosols currently offset 60% of the GHG heating effect, merely stabilizing global S emissions results in substantially greater warming. With vigorous restrictions on emissions of all GHGs (and ozone precursors), there is still greater warming throughout the 21st century than if no restrictions are placed on GHG or S emissions. The good news is that, if reductions in S emissions are regarded as inevitable, then the extra climate warming can be largely nullified (in the global mean) by the end of the 21st century. Furthermore, temperatures have stabilized by that time for the scenario in which restrictions are placed on S and all GHG emissions, but continue to rise for the scenario with no emission restrictions (and would rise rapidly once S emissions decrease, as they must do eventually).

One could extend this analysis further to examine regional and seasonal temperature changes using models of intermediate complexity. Indeed, this is a case where analysis at the large regional scale is highly appropriate in spite of uncertainties in the regional effects for GHG increases alone, simply because the aerosol cooling effect is highly concentrated regionally. Nevertheless, the examples presented here serve to illustrate some of the tradeoffs involving multiple GHGs and S emissions.

11.4 Inverse calculations using simple models

The alternative to specifying emission scenarios and working forward to calculate GHG concentrations and thence climatic change, is to specify a variety of concentration pathways corresponding to stabilization at different concentrations, and to work out the permitted emissions. Past emissions and permitted future CO_2 emissions can be decomposed into the product of the following factors:

- population (P),
- economic output (dollars) per person ($/P),
- average energy consumption (joules) per dollar of economic output (J/$),
- average CO_2 emission (kg) per joule of energy consumption (kg/J).

Thus,

$$\text{Emission} = P \times (\$/P) \times (J/\$) \times (kg/J) \qquad (11.1)$$

This equation makes it clear that future emissions depend on the future human population, the growth of per capita income, the kinds of economic activities undertaken and the efficiency with which energy is used (which together determine J/$), and the mix of energy sources used and the efficiency with which they are transformed into intermediate energy forms such as electricity and refined petroleum products, which together determine kg/J. Hoffert *et al.* (1998) performed this decomposition for the IPCC BAU scenario (IS92a), and the results are presented in Figure 11.6. The BAU scenario assumes that the human population reaches 11.3 billion by 2100 (about double the present population) and that the average per capita income increases by a factor of 6, giving an increase in size of the global economy by a factor of 12. CO_2 emissions would grow by a comparable factor, were it not for the fact that the energy intensity of the global economy is assumed to decrease by 1% per year throughout the 21st century (each year, 1% less energy is required to produce a dollar of economic output than the year before). This is comparable to the rate of decrease that has occurred over the

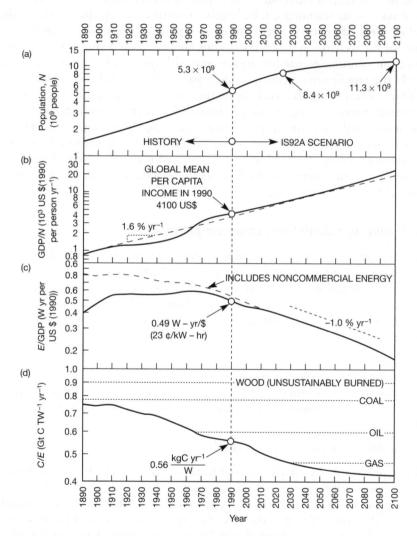

Figure 11.6 Decomposition of past global fossil fuel carbon emissions, and future emissions under the IS92a scenario, according to (a) global population, (b) per capita income, (c) primary energy use per unit of GDP, and (d) carbon emission per unit of primary energy. The horizontal lines in (d) give the carbon intensity for individual fuels. Redrafted from Hoffert et al. (1998).

past century, and reflects both increases in the efficiency with which energy is used, and a shift from energy-intensive primary industries to less energy-intensive service industries. Finally, the carbon intensity of the energy supply system is assumed to decrease by about 25% as a result of an increase in the proportion of non-carbon energy sources (nuclear power and/or renewable energy).

Figure 11.7 shows the concentration stabilization pathways for CO_2 of Wigley *et al.* (1996), with concentrations stabilizing at 450, 550, 650, and 750 ppmv. Figure 11.8(a) shows the permitted global fossil fuel CO_2 emissions, as computed using the carbon cycle model of Jain *et al.* (1995) and presented in Hoffert *et al.* (1998). Also given is the IPCC BAU scenario (the IS92a scenario), which is the same BAU scenario as used previously in this chapter. Stabilization of CO_2 at a concentration of 450 ppmv would still result in the equivalent of close to a CO_2 doubling from the pre-industrial concentration of 280 ppmv, given the radiative heating effect of increases in other GHGs that would also occur. Stabilization at 450 ppmv requires reducing emissions to about half the current emission rate by 2100. Also shown in Figure 11.8(a) is the breakdown of emissions into emissions from natural gas, oil, and coal; the fuels are assumed to be used in this order, subject to the cap on total emissions (recall from Table 3.1 that the CO_2 emission per unit of fossil fuel energy is smallest for natural gas, followed by oil and then by coal). Figure 11.8(b) shows the provision of primary power corresponding to each of the scenarios, while Figure 11.8(c) shows the amount of carbon-free power that is required. This is given by the difference between the primary power required for the IS92a scenario and the total fossil fuel power that is permitted for the stabilization scenarios. The permitted fossil fuel primary power for the stabilization scenarios is computed by assuming that the same amount of natural gas and oil is used as for the IS92a scenario, and any remaining permitted emission is assigned to coal. Even for the BAU scenario, the required non-fossil energy

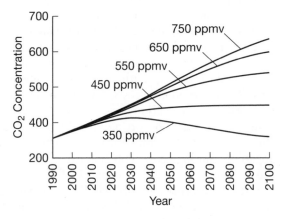

Figure 11.7 The CO_2 stabilization scenarios of Wigley *et al.* (1996). The concentrations at which CO_2 eventually stabilizes are shown. Produced based on data kindly provided in electronic form.

supply reaches 11 TW (terawatts, $1\,TW = 10^{12}\,W$), which is equal to the total global energy supply in 1990. For the stabilization scenarios, an even larger non-fossil power supply is needed.

These results indicate that a significant expansion of non-fossil fuel energy supply – either nuclear energy or renewable energy – is implicit even in BAU scenarios, scenarios in which atmospheric CO_2 reaches several times the pre-industrial concentration. The required non-fossil energy supply can be reduced if the energy intensity of the global economy can be made to decrease at a faster rate than 1% per year. The tradeoff between the required non-fossil energy supply and the rate of decrease of energy intensity is shown in Figure 11.9 for stabilization of CO_2 at 550 ppmv. More of one allows less of the other. If the energy intensity were to improve by 2% per year rather than 1% per year, almost no expansion of non-fossil energy supply would be required.

Since the ultimate objective of the UNFCCC will almost certainly require *some* limitation of CO_2 emissions to below the IS92a scenario, there is already a clear policy implication, even in the absence of detailed scientific information: a major effort is required to ensure that non-fossil energy supply equal to at least the current total global energy supply is available a mere 50 years from now, or a major effort is required to accelerate the improvement in the efficiency with which energy is used. Since implementation of either of these options, to the extent required to stabilize atmospheric CO_2 at 550 ppmv, would be very difficult, and there could be further, unexpected difficulties, one can go further and state that strong efforts are *simultaneously* required on both fronts. This is an example of robust policy advice that can be derived based on the first-order relationships between the driving factors and climate change – relationships that are most clearly illustrated with the aid of simple models.

11.5 Considerations of risk and surprise

The development of a policy response to the threat of global warming should take into account the possibility of unpleasant surprises. Nature is likely to provide many "surprises" – abrupt and largely unanticipated changes in the climate system. The greater the global average warming that is allowed to occur, the more likely it would seem that surprises will occur. True surprises, by definition, cannot be predicted in advance. Surprises are low-probability events with high cost (in terms of either economic costs, loss of human life, or damage to ecosystems). If one were to plot various possible "costs" of global warming versus the probability of that cost occurring in reality, the result would be a highly skewed distribution, with a long tail to the right representing events with very low probability but high cost. This is illustrated in Figure 11.10. The cost of the most likely outcome – given by the peak of the probability distribution function (point x_1) – is substantially less than the average cost of the whole range of possible outcomes. This is because the low-probability, high-impact costs disproportionately affect the average cost, and are not offset by low-probability, low-impact costs (there being no tail to the left in Figure 11.10). The extent to which one considers low-probability, high-cost outcomes

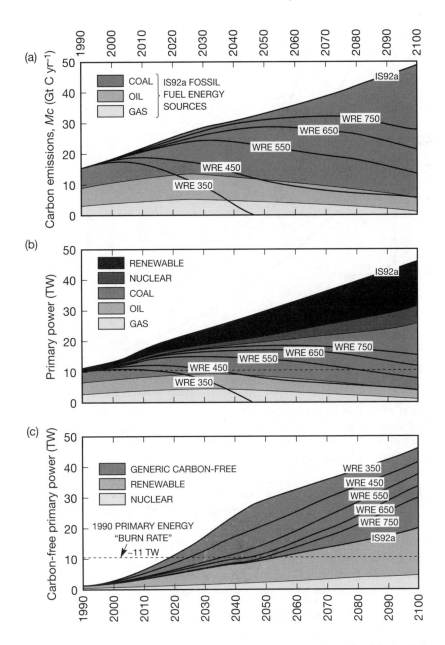

Figure 11.8 (a) Fossil fuel emissions for the IS92a scenario, and premitted fossil fuel emissions for the CO_2 concentration stabilization pathways shown in Figure 11.7. Also shown are the contributions of natural gas, oil, and coal to the permitted CO_2 emissions. (b) The total primary power requirement, and (c) the carbon-free primary power requirement, corresponding to the IS92a and CO_2 stabilization scenarios. Redrafted from Hoffert *et al.* (1998).

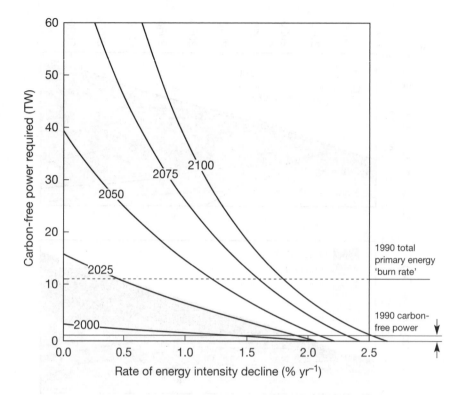

Figure 11.9 The tradeoff between the amount of carbon-free power in various years and the rate of improvement in the energy intensity of the global economy (J/$) required in order to stabilize the CO_2 concentration at twice the pre-industrial value. Redrafted from Hoffert *et al.* (1998).

in developing a policy response depends on how risk-averse one is. This will depend in part on one's viewpoint concerning our obligations to future generations and the sense of stewardship that we hold (see Brown, 1992, for a discussion of concepts of stewardship and fiduciary trust in the context of the global warming issue).

Among the possible sources of "surprises" in the climate system response to GHG emissions are the following:

- an abrupt reorganization of the ocean circulation, something that is permitted according to very basic principles and is frequently found in model simulations (Broecker, 1997);
- an abrupt disappearance of permanent sea ice in the Arctic due to a breakdown of the underlying salinity structure, which at present isolates the surface water from relatively warm water at intermediate depths of North Atlantic origin (Aagaard *et al.*, 1981);
- an abrupt increase in climate sensitivity, leading to a period of rapid (and hence disruptive) climatic change (Section 9.1);

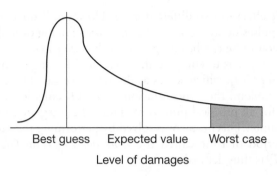

Figure 11.10 Schematic illustration of the probability distribution of the future costs of anthropogenic climatic change. Reproduced from Rothman (1999).

- abrupt changes in precipitation patterns, something that has occurred repeatedly in the past (e.g., Gasse and van Campo, 1994) and could very well be triggered by global-scale climatic changes; and
- an abrupt dieback of forests, as climatic zones shift faster than the rate at which forest zones can migrate, leading to critical thresholds suddenly being surpassed (Section 7.2).

The climate system is sufficiently complex, climate models are sufficiently simple, and our understanding of the phenomenon of global warming and climatic change sufficiently rudimentary, that surprises of some sort can certainly be expected. This in turn calls for a precautionary approach in developing a policy response to the prospect of significant, rapid, and essentially irreversible climatic change.

11.6 Global Warming Potential

The final topic under the subject of the science–policy interface that will be discussed here is that of the "Global Warming Potential" (GWP). The GWP represents an attempt to develop a single index for quantitatively comparing the climatic effects of equal emissions of different GHGs. The GWP is an attempt to simplify complex scientific information for purposes of policy analysis. However, as discussed below, the resultant index cannot be scientifically defended, and it is not necessary, given the ease with which climatic changes with different emission scenarios can be calculated (as exemplified by the examples given in Section 11.3).

As noted in Chapter 2 (Table 2.2), different GHGs differ in the radiative forcing per molecule and in the average lifespan of their molecules in the atmosphere. Thus, if equal quantities of two different gases are emitted at the same time, the relative amounts of the two gases will change over time. Hence, the relative radiative forcings – given by the ratio of the forcing per molecule times the number of molecules remaining in the atmosphere for the two gases – will also change over time, so it is not possible to uniquely specify the relative climatic effect of two different gases. However, if the concentration of both emission pulses decays to zero, the *integrated* forcing over all time arising from the emission pulse can be compared for the two gases.

The natural choice is to compare the integrated forcing of the gas in question with that for CO_2, which serves as the reference gas. There is, however, a fundamental problem: the concentration of a CO_2 emission pulse does not decay to zero (for all practical purposes). Thus, the integrated forcing over all time will be infinite. This problem is solved by performing the integration over some finite period of time called the *time horizon*. The *Global Warming Potential (GWP)* is thus defined as

$$\text{GWP}_i\,(T) = \frac{\int_0^T f_i(t)\,C_i(t)\,dt}{\int_0^T f_r(t)\,C_r(t)\,dt} \tag{11.2}$$

where $f_i(t)$ and $f_r(t)$ are the heat-trapping abilities per unit mass, $C_i(t)$ and $C_r(t)$ are the amounts remaining in the atmosphere at time t, the subscripts i and r refer to the gas in question and to the reference gas, respectively, and T is the time horizon over which the integration is performed. The major drawback of the GWP is that it depends on the choice of T: for gases that are removed more quickly than CO_2, the GWP will be smaller the longer the time horizon, while for gases that are removed more slowly than the initial rate of removal of CO_2, the GWP will be larger the longer the time horizon. Whatever time horizon is chosen for the GWP, the computed GWP will not accurately reflect the relative warming effects of different gases for other time horizons.

The Kyoto Protocol places restrictions on a basket of six gases or groups of gases: CO_2, CH_4 (methane), N_2O (nitrous oxide), SF_6 (sulphur hexafluoride), the HFCs (hydrofluorocarbons), and the PFCs (perfluorocarbons). The Kyoto Protocol will use GWPs as computed by the Intergovernmental Panel on Climate Change (IPCC) for a time horizon of 100 years.

There are a number of significant uncertainties associated with GWPs, that are more fully reviewed in Harvey (1993): (i) the direct radiative forcing of some gases (the HFCs in particular) is still uncertain by up to ±30%; (ii) some gases (the halocarbons and methane in particular) have important indirect radiative forcings that are very difficult to compute accurately; (iii) atmospheric lifetimes of some of the gases being compared with CO_2 are still uncertain and, in the case of CH_4, are expected to change over time; and (iv) changes in climate could cause important changes in the net rate of removal of CO_2 from the atmosphere, and this would alter the computed GWPs for all gases. Other uncertainties, such as the assumed future concentrations of the GHGs, which alter the $f_i(t)$ and $f_r(t)$, are less important. The use of the GWP as defined above contains a number of implicit assumptions as well. Most importantly, it is assumed that the climatic effect of a given GHG is directly proportional to its radiative forcing (that is, that the climate sensitivity is the same for all gases). As discussed in Section 6.6, this assumption is valid to within ±20% when comparing the GHGs that are covered by the Kyoto Protocol. The concept of a GWP is completely inapplicable to gases such as CO and NO_x (which affect climate through their role in the production of tropospheric ozone) and to aerosols, since the climate sensitivity to ozone and aerosol radiative forcing appears to be quite different from that for well-mixed GHGs.

A much less problematic application of GWPs is in comparing the impact on future climatic change of alternative replacements for CFC-11 and CFC-12. The CFCs are powerful GHGs but are being phased out because of their impact on stratospheric ozone. HFC replacements have no effect on the ozone layer but most are powerful GHGs. In this case, CFC-11 can be used as a reference gas and the time horizon T can be taken to infinity. The difficulties involving CO_2 as a reference gas can thus be avoided. One may still choose to carry out the integration in Equation (11.2) over a finite time horizon rather than over all time, if one is interested in near-time rates of warming rather than the integrated warming. Calculated values of GWPs using CFC-11 as a reference gas can be found in Table 2.8 of Schimel *et al.* (1996).

To summarize, the concept of GWP is an attempt to force the complexities of nature into a single number. This is not scientifically justifiable. Even if the heating effects and lifetimes of GHGs were accurately known, the GWP calculated for any one time horizon will not accurately reflect the relative climatic effects of gases for other time horizons. It is more scientifically justifiable to compare climatic changes, and rates of climatic change, for alternative emission scenarios using simple models that can perform forward calculations. In this way, alternative ways of achieving overall targets concerning maximum permitted climatic change can be compared. In addition, attention can be directed to the question of offsetting the inevitable acceleration of warming as sulphur emissions are reduced (as discussed in this chapter), rather than the question of how much reduction in CO_2 emissions can be avoided through reduced emissions of other gases. However, the political process has demanded a single number to intercompare different GHGs. This is because current international agreements have ignored scientific reality by focusing on a "basket" of GHGs rather than by framing gas-by-gas restrictions. The latter would eliminate the most problematic applications of GWPs altogether.

Questions for further thought

1. In simulating past observed temperature trends, why does a larger assumed climate sensitivity require a larger aerosol cooling effect?
2. In simulating future atmospheric CO_2 concentrations for a given emission scenario, carbon cycle models are first adjusted so as to simulate the observed CO_2 buildup so far. In constraining carbon cycle models to match the observed CO_2 buildup, why does a larger assumed current rate of emission due to deforestation result in a smaller projected future CO_2 concentration?
3. Under what conditions could a decrease in CO_2 emissions initially cause an *acceleration* of global warming?

Glossary

Acid rain Rainfall with a greater than average acidity (lower pH) due to the presence of acids produced from anthropogenic emissions of sulphur and nitrogen oxides.

Adjustment time The time required for the perturbation in the amount of a gas in the atmosphere to decrease to $1/e$ of the amount initially injected, where the initial injection occurred as a sudden pulse. This will be the same as the lifespan or turnover time τ if the presence of the pulse does not increase the lifespan of the pre-existing gas.

Aerosol Small airborne particles.

Albedo The fraction of solar radiation incident on a surface that is reflected.

Anthropogenic Of human origin.

Asymbiotic Used here to refer to nitrogen-fixing bacteria that are not living in a symbiotic relationship with a higher plant, but rather live freely in the soil. See also *symbiotic*.

Bathymetry Refers to the shape of an ocean basin, or the variation of depth with horizontal position.

Biological pump The process whereby carbon and nutrients are transferred from the surface waters to the deep ocean, due to the occurrence of photosynthesis near the ocean surface, the settling of some dead organic matter into the subsurface layer, and the decomposition of organic matter in deeper water.

Biomass Refers to living plant or animal matter.

Biosphere Refers to all living organisms – the living "sphere".

Black carbon Refers to soot aerosol – an aerosol consisting of highly absorbing carbon compounds that are produced as a result of the incomplete combustion of fossil fuels or biomass.

Brine A salt-rich solution produced by the ejection of salt during the freezing of sea water to produce sea ice.

Buffer factor The ratio of the percentage increase in the partial pressure of CO_2 in a layer of the ocean to the percentage increase in total dissolved carbon in that layer.

C_3 plant A plant that uses one of the two major pathways for photosynthesis, in which intermediate compounds with three carbon atoms are produced.

C_4 plant A plant that uses one of the two major pathways for photosynthesis, in which intermediate compounds with four carbon atoms are produced.

Calcium carbonate The material ($CaCO_3$) used to construct the skeletons in several major groups of marine micro-organisms.

Calving The breaking off into the ocean water of pieces of an ice sheet.

Carbonate (CO_3^{2-}) One of three forms in which inorganic carbon occurs in dissolved form in ocean water, the other two forms being as dissolved CO_2 and as bicarbonate (HCO_3^-).

CFCs Chlorofluorocarbons – a group of compounds built around one or more carbon atoms, and containing chlorine and fluorine.

Clausius–Clapeyron equation The concave upward relationship between the vapour pressure of water at saturation and temperature.

Climate sensitivity The ratio of the global mean equilibrium or steady-state temperature change to the radiative forcing that caused the temperature change, taking into account only the fast feedbacks.

CO_2 fertilization Refers to the tendency for higher atmospheric CO_2 concentrations to stimulate plant growth. This can occur as a result of the direct stimulation of photosynthesis, or an improvement in the efficiency of water use, which allows more plant growth for a given supply of water.

Convection The processes whereby parcels of air (in the atmosphere) or water (in the ocean) are exchanged vertically when the underlying air parcel becomes less dense than the overlying air due to heating, or the overlying water parcel becomes more dense than the underlying water due to cooling or salinification of the overlying parcel.

Coriolis force A force that varies in direct proportion to the speed of an object, and which acts perpendicularly to the right (left) of the motion in the NH (SH).

Cryosphere Refers to the cold sphere – glaciers, ice sheets, sea ice, and land snow cover.

Diapycnal mixing Mixing in the ocean in a direction perpendicular to isopycnal (constant-density) surfaces.

Diffusive mixing Refers to the net mixing due to small-scale eddies and turbulence. See also *diapycnal mixing* and *isopycnal mixing*.

Dissolved inorganic carbon The sum of dissolved CO_2, carbonate, and bicarbonate in the ocean.

Downregulation The tendency for the initial stimulation of photosynthesis caused by higher atmospheric CO_2 concentration to decrease over time.

Effective climate sensitivity The climate sensitivity that would occur in steady state if the strengths of the various feedback processes at a given point in the transient response were to persist.

El Niño A particular anomaly in the temperature and pressure patterns in the tropical Pacific Ocean, with worldwide effects, in which the eastern tropical Pacific is warmer than average and has lower surface pressure than average, while opposite changes occur in the western tropical Pacific.

Emissivity The ratio of the amount of radiation emitted by an object to the maximum permissible emission (the latter being given by the Stefan–Boltzmann relationship).

Equilibrium climatic change The climatic change that occurs after the atmosphere and ocean have fully adjusted to the imposed radiative forcing.

Equivalent CO_2 increase The increase in CO_2 concentration alone that has the same global mean radiative forcing as that arising from increases in a variety of well-mixed greenhouse gases.

Evapotranspiration The combination of evaporation of water directly from the surface of plants, and the transpiration or transfer of water from the interior of leaves to the exterior, largely through small openings called stomata.

External forcing The radiative forcing that results from some perturbation that is external to the climatic system, such as a change in solar luminosity, volcanic eruptions, or anthropogenic emissions of GHGs or aerosol precursors.

Extratropical cyclone A travelling synoptic-scale (1000 km) low-pressure system that occurs in middle and high latitudes, bringing with it rain.

Faculae Bright areas on the sun's surface, usually occurring in close proximity to sunspots. The presence of faculae contributes to an increase in the energy output from the sun.

Fast feedbacks Refers to those feedbacks in the climate system that operate over a period of days to months, and are used in the computation of climate sensitivity.

Flux A rate of flow of something.

General Circulation Model (GCM) A three-dimensional model of the atmosphere or ocean in which the motions at the scale of the model grid are simulated (along with many other processes).

Greenhouse gas Any gas that emits and absorbs radiation in the infrared part of the electromagnetic spectrum, thereby tending to make the climate warmer than it would be in its absence.

Grounding line The boundary between the portion of an ice sheet that is on bedrock below sea level and the portion that is floating. Applicable to the West Antarctic Ice Sheet.

Growth respiration Respiration required to power the process of growth in a plant.

Halocarbons Compounds containing either chlorine, bromine, or fluorine, together with carbon. Such compounds can act as powerful greenhouse gases in the atmosphere. The chlorine- and bromine-containing halocarbons are also involved in depletion of the ozone layer.

Heterotrophic respiration Respiration by the animals within an ecosystem.

Hydrochlorofluorocarbons (HCFCs) Compounds similar to the CFCs, except that one or more chlorine or fluorine atoms have been replaced with a hydrogen atom, thereby making them decompose much more readily.

Hydrofluorocarbons (HFCs) Carbon- and fluorine-containing compounds which, by virtue of the absence of chlorine or bromine, have no direct effect on stratospheric ozone, but do act as greenhouse gases.

Immobilization The process whereby nutrients that are released by mineralization are taken up by microbes and become unavailable for use by plants.

Impulse response The variation over time in the amount of a gas remaining in the atmosphere following a sudden injection.

Infrared radiation Electromagnetic radiation between wavelengths of $4.0\,\mu m$ and $50.0\,\mu m$, which is emitted by objects at temperatures typical of the Earth's surface and atmosphere.

Inverse problem Here, this refers to the problem of determining the magnitude and/or geographical distribution of the sources and sinks of a given gas based on the observed geographical and/or temporal variation in its concentration in the atmosphere.

Inversion A condition where temperature increases with increasing height in the atmosphere, which is opposite to the usual situation of decreasing temperature with increasing height.

Isopycnal mixing Mixing in the ocean along surfaces of constant density.

Isopycnal surface A set of adjoining points in 3-D space having the same density – a constant-density surface.

Isotope A variety of an element having a different number of neutrons in its nucleus and hence a different atomic mass.

Kyoto Protocol The agreement reached in Kyoto, Japan, on 10 December 1997 to limit emission of greenhouse gas by industrialized countries.

Lapse rate The negative of the rate of change of temperature with height. Since temperature normally decreases with increasing height, this gives a positive lapse rate. A greater lapse rate means a greater rate of decrease of temperature with increasing height.

Latent heat The heat that is required to melt ice or evaporate water, and that is released when freezing or condensation occurs.

Lithosphere The crust or outer layer of the Earth.

Maintenance respiration The background rate of respiration required to maintain living processes. See also *growth respiration* and *photorespiration*.

Mineralization The process whereby nutrients in organic matter are converted back to forms that can be readily used by plants or micro-organisms.

Mixed layer The surface layer of the ocean, that is well mixed by the action of wind-induced turbulence or the sinking of water parcels that become denser due either to surface cooling or to an excess of evaporation over precipitation plus runoff.

Monsoon The seasonal shift in wind patterns over south Asia and west Africa, whereby air flows from ocean to continent in summer and the reverse in winter. The summer monsoon brings rainfall to the affected regions.

Negative feedback A feedback in which the initial change (ΔA) provokes a change in some intermediate quantity (ΔB), and the change ΔB acting alone tends to provoke a further change in A that is opposite to the initial change. A negative feedback has a stabilizing effect.

Net Primary Production (NPP) The rate of photosynthesis by a plant minus the rate of growth and maintenance respiration.

Net radiation Absorbed solar radiation minus emitted infrared radiation.

Nitrification The process that converts NH_3 or NH_4^+ into NO_2^- or NO_3^-.

Nitrogen fixation The process that converts atmospheric N_2 into NH_3 or NH_4^+.

Nutrient Utilization Efficiency (NUE) An increase in the C:nutrient ratio in plant organic matter, such that less nutrients are required per unit of assimilated carbon. This can occur as a consequence of the biochemical downregulation of the photosynthetic response to higher CO_2, in which the concentration of the rubisco enzyme decreases.

Oxidation An element is oxidized when it gives up electrons to an electron acceptor or oxidizing agent, the most common electron acceptor being oxygen. It so doing it becomes chemically bound to the oxidizing agent. Biomass and fossil fuels are oxidized to produce CO_2 and H_2O.

Parameterization The technique of representing processes in a model that cannot be explicitly resolved at the resolution of the model (sub-grid scale processes) by relationships between the area-averaged effect of such sub-grid scale processes and the larger-scale variables (that are represented in the model).

Partial pressure The pressure due to a given gas acting alone.

Photorespiration Respiration that is mediated by the same enzyme that catalyses photosynthesis (the rubisco enzyme) and that occurs only in the presence of sunlight. When an O_2 molecule attaches to the rubisco enzyme, photorespiration can occur, whereas when a CO_2 molecule attaches, photosynthesis can occur.

Planetary boundary layer The atmospheric layer next to the Earth's surface, in which there is intense vertical mixing.

Polar amplification Refers to the tendency, in computer climate models and in paleoclimatic data, for temperature changes at high latitudes to be greater than temperature changes at low latitudes when there is an overall warming or cooling of the climate.

Positive feedback A feedback in which the initial change (ΔA) provokes a change in some intermediate quantity (ΔB), and the change ΔB acting alone tends to provoke a further change in A that is in the same direction as the initial change. A positive feedback has a destabilizing effect.

Potential evapotranspiration The rate of evapotranspiration that would occur from a wet surface, or from a well-watered plant.

Precursor A chemical that is emitted into the atmosphere and reacts with other chemicals to form a compound of interest. For example, CO, hydrocarbons, and NO_x are all precursors of tropospheric ozone, while SO_2 is a precursor for sulphate aerosol.

Radiative damping The increase in net emission of radiation to space in response to the warming of an object.

Radiative forcing The perturbation in the radiative balance at the tropopause in response to some externally applied perturbation (such as an increase in greenhouse gas concentrations other than water vapour).

Residence time The average length of time that a molecule spends in a reservoir before being removed.

Respiration The process of oxidizing or "burning" organic matter to provide energy, usually through the consumption of O_2. CO_2 is produced as a by-product of respiration. See also *growth respiration, heterotrophic respiration, maintenance respiration,* and *photorespiration.*

Senescence The process of ageing in a plant, characterized by a slowing down of physiological activity, decreasing growth rates, and increased sensitivity to abiotic and biotic stresses. This occurs at both the decadal and the seasonal time scales in deciduous trees, and is associated with leaf drop during autumn.

Sensible heat Heat that can be sensed or felt, related to the temperature of an object, gas, or fluid.

Sink A reservoir into which a gas is absorbed and stored for a long period of time.

Stomata Small openings on the surfaces of leaves that permit the passage of CO_2 into the interior of the leaf and the passage of water vapour from inside to outside the leaf.

Stomatal conductance The ease with which gases pass through the stomata. A greater conductance implies a smaller resistance.

Stratosphere The layer of the atmosphere between a height of 10–17 and 50 km which contains the ozone "layer", and in which temperatures are either uniform with height (the lower stratosphere) or increase with increasing height (producing an inversion).

Sulphur hexafluoride (SF_6) An extremely powerful greenhouse gas used in the aluminium industry.

Sunspot Dark areas seen on the sun's surface, in some cases perceptible with the naked eye. The presence of sunspots contributes to a reduction in the energy output from the sun. See also *faculae.*

Surprise-free scenario Refers to the fact that the way the future unfolds in reality will almost certainly contain unexpected outcomes or "surprises" which, by definition, cannot be anticipated in the scenarios that are built using models. Such model-based projections therefore constitute surprise-free scenarios.

Symbiotic A mutually beneficial relationship between two different species.

System A set of components that interact with each other.

Thermohaline overturning The vertical overturning circulation in the oceans that is driven by horizontal differences in the density of water.

Transient climatic change The time-dependent climatic change during the transition from one equilibrium or statistically steady-state climate to another.

Tropopause The boundary between the troposphere and the stratosphere.

Troposphere The lowest layer of the atmosphere, in which weather disturbances occur. The troposphere varies in thickness from about 20 km at the equator to about 14 km at the poles.

Upwelling–diffusion model (UD model) A one-dimensional ocean model that is averaged horizontally (globally) but resolved vertically, and in which vertical mixing by diffusion and the upwelling of the thermohaline circulation are represented.

Water use efficiency (WUE) The ratio of carbon assimilation by photosynthesis to water loss through transpiration. WUE efficiency tends to increase as the atmospheric CO_2 concentration increases.

Well-mixed gases Gases whose lifetime in the atmosphere is sufficiently long (several years or more) that they have time to spread throughout the atmosphere before being removed. As result, they acquire a fairly uniform concentration within the atmosphere, the concentration being more uniform the longer the atmospheric lifetime and the more dispersed the sources and sinks.

References

Aagaard, K., Coachman, L.K. and Carmack, E. (1981) On the halocline of the Arctic Ocean. *Deep Sea Research*, **28A**, 529–545.

Andreae, M.O. (1995) Climatic effect of changing atmospheric aerosol levels, in Henderson-Sellers, A. (ed.), *Future Climates of the World: a Modelling Perspective, World Survey of Climatology*, Volume 16, Elsevier, Amsterdam, 347–398.

Angell, J.K. (1997) Annual and Seasonal Global Temperature Anomalies in the Troposphere and Low Stratosphere, 1958–1996, Carbon Dioxide Information Analysis Center, Report NDP–008, Oak Ridge National Laboratory, Oak Ridge, Tennessee.

Antoine, D. and Morel, A. (1995) Modelling the seasonal course of the upper ocean pCO_2 (i). Development of a one-dimensional model. *Tellus*, **47B**, 103–121.

Archer, D. (1996a) An atlas of the distribution of calcium carbonate in sediments of the deep sea. *Global Biogeochemical Cycles*, **10**, 159–174.

Archer, D. (1996b) A data-driven model of the global calcite lysocline. *Global Biogeochemical Cycles*, **10**, 511–526.

Archer, D., Kheshgi, H. and Maier-Reimer, E. (1997) Multiple timescales for neutralization of fossil fuel CO_2. *Geophysical Research Letters*, **24**, 405–408.

Bacastow, R. and Maier-Reimer, E. (1990) Ocean-circulation model of the carbon cycle. *Climate Dynamics*, **4**, 95–125.

Baliunas, S. and Jastrow, R. (1990) Evidence for long term brightness changes of solar-type stars. *Nature*, **348**, 520–523.

Barnes, D.W. and Edmonds, J.A. (1990) *An Evaluation of the Relationship Between the Production and Use of Energy and Atmospheric Methane Emissions*, US Department of Energy, DOE/NBB–088P.

Battle, M. *et al.* (1996) Atmospheric gas concentrations over the past century measured in air from firn at the South Pole. *Nature*, **383**, 231–235.

Bazzaz, F.A. (1990) The response of natural ecosystems to the rising global CO_2 levels. *Annual Review of Ecology and Systematics*, **21**, 167–196.

Beck, L.L., Piccot, S.D. and Kirchgessner, D.A. (1993) Industrial sources, in Khalil, M. (ed.), *Atmospheric Methane: Sources, Sinks, and Role in Global Change*, Springer-Verlag, Berlin, 399–431.

Bekki, S. and Law, K.S. (1997) Sensitivity of the atmospheric CH_4 growth rate to global temperature changes observed from 1980 to 1992. *Tellus*, **49B**, 409–416.

Bekki, S., Law, K.S. and Pyle, J.A. (1994) Effects of ozone depletion on atmospheric CH4 and CO concentrations. *Nature*, **371**, 595–597.

Benkovitz, C.M. *et al.* (1996) Global gridded inventories of anthropogenic emissions of sulfur and nitrogen. *Journal of Geophysical Research*, **101**, 29239–29253.

Berner, R.A. (1994) Geocarb II: A revised model of atmospheric CO_2 over phanero-zoic time. *American Journal of Science*, **294**, 56–91.

Bhaskaran, B., Mitchell, J.F.B., Lavery, J.R. and Lal, M. (1995) Climatic response of the Indian subcontinent to doubled CO_2 concentrations. *International Journal of Climatology*, **15**, 873–892.

Boer, G.J., McFarlane, N.A. and Lazare, M. (1992) Greenhouse gas-induced climate change simulated with the CCC second-generation general circulation model. *Journal of Climate*, **5**, 1045–1077.

Bolker, B.M., Pacala, S.W., Bazzaz, F.A., Canham, C.D. and Levin, S.A. (1995) Species-diversity and ecosystem response to carbon dioxide fertilization – conclusions from a temperature forest model. *Global Change Biology*, **1**, 373–381.

Bonan, G.B. (1997) Effects of land use on the climate of the United States. *Climatic Change*, **37**, 449–486.

Bongaarts, J. (1990) The measurement of unwanted fertility. *Population and Development Review*, **16**, 487–506.

Bongaarts, J. (1994) Population policy options in the developing world. *Science*, **263**, 771–776.

Borzenkova, I.I. (1992) *The Changing Climate During the Cenozoic*, Hydrometeoizdat, Saint Petersburg, 247 pages (in Russian; in English by Kluwer Academic).

Bottomley, M., Folland, C.K., Hsiung, J., Newell, R.E. and Parker, D.E. (1990) *Global Ocean Surface Temperature Atlas (GOSTA)*. Joint Meteorological Office/Massachusetts Institute of Technology Project, HMSO, London, 20 + iv pages and 313 plates.

Boucher, O. and Lohmann, U. (1995) The sulfate–CCN–cloud albedo effect. A sensitivity study with two general circulation models. *Tellus*, **47B**, 281–300.

Branscome, L.E. and Gutowski, W.J. (1992) The impact of doubled CO_2 on the ener-getics and hydrologic processes of mid-latitude transient eddies. *Climate Dynamics*, **8**, 29–37.

Broecker, W.S. (1997) Thermohaline circulation, the Achilles Heel of our climate system: Will man-made CO_2 upset the current balance? *Science*, **278**, 1582–1588.

Brown, P.G. (1992) Climate change and the planetary trust. *Energy Policy*, **20**, 208–222.

Brown, S. *et al.* (1996) Management of forests for mitigation of greenhouse gas emissions, in Watson, R.T., Zinyowera, M.C. and Moss, R.H. (eds), *Climate Change 1995 – Impacts, Adaptations and Mitigation of Climate Change: Scientific–Technical Analysis*, Cambridge University Press, Cambridge, 771–797.

Bryan, K. (1996) The steric component of sea level rise associated with enhance greenhouse warming: a model study. *Climate Dynamics*, **12**, 545–555.

Bryant, E. (1997) *Climate Process and Change*, Cambridge University Press, New York, 209 pages.

Cess, R.D. *et al.* (1993) Uncertainties in carbon dioxide radiative forcing in atmos-pheric general circulation models. *Science*, **262**, 1252–1255.

Chan, J.C.L. and Shi, J. (1996) Long-term trends and interannual variability in tropical cyclone activity over the western North Pacific. *Geophysical Research Letters*, **23**, 2765–2767.

Charlson, R.J., Langner, J., Rodhe, H., Leovy, C.B. and Warren, S.G. (1991) Perturbation of the Northern Hemisphere radiative balance by backscattering from anthropogenic sulfate aerosols. *Tellus*, **43AB**, 152–163.

Charlson, R.J., Anderson, T.L. and McDuff, R.E. (1992) The sulfur cycle, in Butcher, S.S., Charlson, R.J., Orians, G.H. and Wolfe, G.V. (eds), *Global Biogeochemical Cycles*, Academic Press, London, 285–300.

Chase, T.N., Pielke, R.A., Kittel, T.G.F., Nemani, R.R. and Running, S.W. (1999) Simulated impacts of historical land cover changes on global climate. *Climate Dynamics* (submitted).

Chen, J. (1998) CO_2 emissions relief through blended cements, www.civil.nwu.edu/ACBM/Chen/CO_2%20emissions%20calculations.htm.

Chin, M., Jacob, D.J., Gardner, G.M., Foreman-Fowler, M.S. and Spiro, P.A. (1996) A global three-dimensional model of tropospheric sulfate. *Journal of Geophysical Research*, **101**, 18667–18690.

Christidis, N., Hurley, M.D., Pinnock, M.D., Shine, K.P. and Wallington, T.J. (1997) Radiative forcing of climate change by CFC-11 and possible CFC replacements. *Journal of Geophysical Research*, **102**, 19597–19609.

Christopherson, R.W. (1992) *Geosystems: An Introduction to Physical Geography*, Macmillan, New York, 663 pages + appendices.

Christy, J.R., Spencer, R.W. and Lobl, E.S. (1998) Analysis of the merging procedure for the MSU daily temperature time series. *Journal of Climate*, **11**, 2016–2041.

Ciais, P. *et al.* (1995) Partitioning of ocean and land uptake of CO_2 as inferred by $\delta^{13}C$ measurements from the NOAA Climate Monitoring and Diagnostics Laboratory global air sampling network. *Journal of Geophysical Research*, **100**, 5051–5070.

Colman, R.A. and McAvaney, B.J. (1995) The sensitivity of the climate response of an atmospheric general circulation model to changes in convective parameterization and horizontal resolution. *Journal of Geophysical Research*, **100**, 3155–3172.

Cooke, W.F. and Wilson, J.J.N. (1996) A global black carbon aerosol model. *Journal of Geophysical Research*, **101**, 19395–19409.

Crowley, T.J. (1990) Are there any satisfactory geologic analogs for a future greenhouse warming? *Journal of Climate*, **3**, 1282–1292.

Crowley, T.J. and Baum, S.K. (1995) Is the Greenland ice sheet bistable? *Paleoceanography*, **10**, 357–363.

Crutzen, P.J. and Andreae, M.O. (1990) Biomass burning in the tropics: impact on atmospheric chemistry and biogeochemical cycles. *Science*, **250**, 1669–1678.

Crutzen, P.J. and Zimmermann, P.H. (1991) The changing photochemistry of the troposphere. *Tellus*, **43AB**, 136–151.

Cuffey, K.M. and Clow, G.D. (1997) Temperature, accumulation, and ice sheet elevation in central Greenland through the last deglacial transition. *Journal of Geophysical Research*, **102**, 26383–26396.

Dai, A. and Fung, I.Y. (1993) Can climate variability contribute to the "missing" CO_2 sink? *Global Biogeochemical Cycles*, **7**, 599–609.

de la Mare, W.K. (1997) Abrupt mid-twentieth-century decline in Antarctic sea-ice extent from whaling records. *Nature*, **389**, 57–60.

De Wolde, J.R., Bintanja, R. and Oerlemans, J. (1995) On thermal expansion over the last one hundred years. *Journal of Climate*, **8**, 2881–2891.

De Wolde, J.R., Juybrechts, P., Oerlemans, J. and van de Wal, R.S.W. (1997) Projections of global mean sea level rise calculated with a 2D energy-balance climate model and dynamic ice sheet models. *Tellus*, **49A**, 486–502.

Del Genio, A.D., Kovari, W. and Yao, M.-S. (1994) Climatic implications of the seasonal variation of upper tropospheric water vapor. *Geophysical Research Letters*, **21**, 2701–2704.

Del Genio, A.D., Yao, M.-S., Kovari, W. and Lo, K.K.-W. (1996) A prognostic cloud water parameterization for global climate models. *Journal of Climate*, **9**, 270–304.

DeMaria, M. and Kaplan, J. (1994) Sea surface temperature and the maximum intensity of Atlantic tropical cyclones. *Journal of Climate*, **7**, 1324–1334.

Demeritt, D. and Rothman, D.D. (1998a) Comments on J.B. Smith (*Climatic Change* 32, 313–326) and the aggregation of climate damage costs. *Climatic Change*, 40, 699–704.

Demeritt, D. and Rothman, D.S. (1998b) Figuring the costs of climate change: an assessment and critique. *Environment and Planning A* (in press).

Dey, B. and Bhanu Kumar, O.S.R.U. (1982) An apparent relationship between Eurasian spring snow cover and the advance period of the Indian summer monsoon. *Journal of Applied Meteorology*, 21, 1929–1932.

Diaz, S., Grime, J.P., Harris, J. and McPherson, E. (1993) Evidence of a feedback mechanism limiting plant response to elevated carbon dioxide. *Nature*, 364, 616–617.

Dickson, R.R. (1984) Eurasian snow cover versus Indian monsoon rainfall – an extension of Hahn–Shukla results. *Journal of Climate and Applied Meteorology*, 23, 171–173.

Dixon, R.K., Brown, S., Houghton, R.A., Solomon, A.M., Trexler, M.C. and Wisniewski, J. (1994) Carbon pools and flux of global forest ecosystems. *Science*, 263, 185–190.

Dlugokencky, E.J., Masarie, K.A., Lang, P.M. and Tans, P.P. (1998) Continuing decline in the growth rate of the atmospheric methane burden. *Nature*, 393, 447–450.

DOE (US Department of Energy) (1992) Electric Power Annual 1990, DOE/EIA-0348(90), Washington, DC.

Douglas, B.C. (1991) Global sea level rise. *Journal of Geophysical Research*, 96, 6981–6992.

Douglas, B.C. (1992) Global sea level acceleration. *Journal of Geophysical Research*, 97, 12699–12706.

Douglas, B.C. (1995) Global sea level change: determination and interpretation. *Reviews of Geophysics* (supplement), 1425–1432.

Douglas, B.C. (1997) Global sea rise: a redetermination. *Surveys in Geophysics*, 18, 279–292.

Drake, B.G., Gonzàlez-Meler, M.A. and Long, S.P. (1996) More efficient plants: a consequence of rising atmospheric CO_2? *Annual Review of Plant Physiology and Plant Molecular Biology*, 48, 609–639.

Easterling, D.R. *et al.* (1997) Maximum and minimum temperature trends for the globe. *Science*, 277, 364–367.

Ehleringer, J.R., Sage, R.F., Flanagan, L.B. and Pearcy, R.W. (1991) Climate change and the evolution of C_4 photosynthesis. *Trends in Ecology and Evolution*, 6, 95–99.

EIA (Energy Information Administration) (1998) Appendix B: Characteristics of Coal Supply Contracts, in *Coal Industry Annual 1996* (obtainable from http://www.eia.doe.gov/cneaf/coal/cia/).

Eischeid, J.K., Baker, C.B., Karl, T.R. and Diaz, H.F. (1995) The quality control of long-term climatological data using objective data analysis. *Journal of Applied Meteorology*, 34, 2787–2795.

Eisele, F.L., Mount, G.H., Tanner, D., Jefferson, A., Shetter, R., Harder, J.W. and Williams, E.J. (1997) Understanding the production and interconversion of the hydroxyl radical during the Tropospheric OH Photochemistry Experiment. *Journal of Geophysical Research*, 102, 6457–6465.

Elliott, W.P. and Angell, J.K. (1997) Variations of cloudiness, precipitable water, and relative humidity over the United States: 1973–1993. *Geophysical Research Letters*, 24, 41–44.

EPA (Environmental Protection Agency) (1998) Acid Rain Program – 1997 Compliance Report, Appendix B3 (obtainable from http://www.epa.gov/acidrain/cmprtpt97/cr1997.htm).

Epstein, P.R. *et al.* (1998) Biological and physical signs of climate change: focus on mosquito-borne diseases. *Bulletin of the American Meteorological Society*, **79**, 409–417.

Etheridge, D.M., Pearman, G.I. and Fraser, P.J. (1992) Changes in tropospheric methane between 1841 and 1978 from a high accumulation-rate Antarctic ice core. *Tellus*, **44B**, 282–294.

Etheridge, D.M., Steele, L.P., Langenfelds, X.X., Francey, R.J., Barnola, J.-M. and Morgan, V.I. (1996) Natural and anthropogenic changes in atmospheric CO_2 over the last 1000 years from air in Antarctic ice and firn. *Journal of Geophysical Research*, **101**, 4115–4128.

Evans, J.L. and Allan, R.J. (1992) El Niño/Southern Oscillation modification to the structure of the monsoon and tropical cyclone activity in the Australasian region. *International Journal of Climatology*, **12**, 611–623.

Fan, S., Gloor, M., Mahlman, J., Pacala, S., Sarmiento, J., Takahashi, T. and Tans, P. (1998) A large terrestrial carbon sink in North America implied by atmospheric and oceanic carbon dioxide data and models. *Science*, **282**, 442–446.

Fearnside, P.M. (1995) Hydroelectric dams in the Brazilian Amazon as sources of "greenhouse" gases. *Environmental Conservation*, **22**, 7–19.

Field, C.B., Jackson, R.B. and Mooney, H.A. (1995) Stomatal responses to increased CO_2: implications from the plant to the global scale. *Plant, Cell and Environment*, **18**, 1214–1225.

Fischer, H., Werner, M., Wagenbach, D., Schwager, M., Thorsteinnson, T., Wilhelms, F. and Kipfstuhl, J. (1998) Little ice age clearly recorded in northern Greenland ice cores. *Geophysical Research Letters*, **25**, 1749–1752.

Folland, C.K. and Parker, D.E. (1995) Correction of instrumental biases in historical sea surface temperature data. *Quarterly Journal of the Royal Meteorological Society*, **121**, 319–367.

Folland, C.K., Parker, D.E. and Kates, F.E. (1984) Worldwide marine surface temperature fluctuations 1856–1981. *Nature*, **310**, 670–673.

Forster, P. M. de F. and Shine, K.P. (1997) Radiative forcing and temperature trends from stratospheric ozone changes. *Journal of Geophysical Research*, **102**, 10841–10855.

Forster, P.M. de F., Freckleton, R.S. and Shine, K.P. (1997) On aspects of the concept of radiative forcing. *Climate Dynamics*, **13**, 547–560.

Fraedrich, K., Müller, K. and Kuglin, R. (1992) Northern hemisphere circulation regimes during extremes of the El Niño/Southern Oscillation. *Tellus*, **44A**, 33–40.

Frakes, L.A. (1979) *Climate Through Geologic Time.* Elsevier, Amsterdam, 310 pages.

Francey, R.J., Tans, P.P., Allison, C.E., Enting, I.G., White, J.W.C. and Troller, M. (1995) Changes in oceanic and terrestrial carbon uptake since 1982. *Nature*, **373**, 326–330.

Gaffen, D.J. (1994) Temporal inhomogeneities in radiosonde temperature records. *Journal of Geophysical Research*, **99**, 3667–3676.

Gaffen, D.J., Elliott, W.P. and Robock, A. (1992) Relationships between tropospheric water vapor and surface temperature as observed by radiosondes. *Geophysical Research Letters*, **19**, 1839–1842.

Gas Research Institute (1997) Effect of methane emissions on global warming, Appendix B in *Methane Emissions from the Natural Gas Industry*, Gas Research Institute, Chicago, B2-B28.

Gasse, F. and van Campo, E. (1994) Abrupt post-glacial climate events in West Asia and North Africa monsoon domains. *Earth and Planetary Science Letters*, **126**, 435–456.

Gates, W.L. *et al.* (1996) Climate models – evaluation, in Houghton, J.T., Filho, L.G.F., Callander, B.A., Harris, N., Kattenberg, A. and Maskell, K. (eds), *Climate Change 1995: The Science of Climate Change*, Cambridge University Press, Cambridge, 229–284.

Gifford, R.M. (1992) Interaction of carbon dioxide with growth-limiting environmental factors in vegetation productivity: implications for the global carbon cycle. *Advances in Bioclimatology*, **1**, 24–58.

Gifford, R.M. (1994) The global carbon cycle: a viewpoint on the missing sink. *Australian Journal of Plant Physiology*, **21**, 1–15.

Gordon, H.B., Whetton, P.H., Pittock, A.B., Fowler, A.M. and Haylock, M.R. (1992) Simulated changes in daily rainfall intensity due to the enhanced greenhouse effect: implications for extreme rainfall events. *Climate Dynamics*, **8**, 83–102.

Gorham, E. (1991) Northern peatlands: role in the carbon cycle and probable responses to climatic warming. *Ecological Applications*, **1**, 182–195.

Gornitz, V., Rosenzweig, C. and Hillel, D. (1997) Effects of anthropogenic intervention in the land hydrologic cycle on global sea level rise. *Global and Planetary Change*, **14**, 147–161.

Gough, D.O. (1981) Solar interior structure and luminosity variations. *Solar Physics*, **74**, 21–34.

Gregg, W.W. and Walsh, J.J. (1992) Simulation of the 1979 spring bloom in the mid-Atlantic bight: a coupled physical/biological/optical model. Changes of snow cover, temperature and radiative heat balance over the Northern Hemispher. *Journal of Geophysical Research*, **97**, 5723–5743.

Gregory, J.M. and Oerlemans, J. (1998) Simulated future sea-level rise due to glacier melt based on regionally and seasonally resolved temperature changes. *Nature*, **391**, 474–476.

Groisman, P.Y., Karl, T.R., Knight, R.W. and Stenchikov, G.L. (1994) Changes of snow cover, temperature and radiative heat balance over the Northern Hemisphere. *Journal of Climate*, **7**, 1633–1656.

Grubb, M. (1992) *The Greenhouse Effect: Negotiating Targets*, Royal Institute of International Affairs, London, 56 pages.

Gutzler, D.S. (1992) Climatic variability of temperature and humidity over the tropical Western Pacific. *Geophysical Research Letters*, **19**, 1595–1598.

Haigh, J.D. (1994) The role of stratospheric ozone in modulating the solar radiative forcing of climate. *Nature*, **370**, 544–546.

Hall, N.M.J., Hoskins, B.J., Valdes, P.J. and Senior, C.A. (1994) Storm tracks in a high-resolution GCM with doubled carbon dioxide. *Quarterly Journal of the Royal Meteorological Society*, **120**, 1209–1230.

Hansen, J. and Lebedeff, S. (1987) Global trends of measured surface air temperature. *Journal of Geophysical Research*, **92**, 13345–13372.

Hansen, J., Sato, M. and Ruedy, R. (1995) Long-term changes in the diurnal temperature cycle: implications about mechanisms of global climate change. *Atmospheric Research*, **37**, 175–209.

Hansen, J., Sato, M. and Ruedy, R. (1997a) Radiative forcing and climate response. *Journal of Geophysical Research*, **102**, 6831–6864.

Hansen, J. *et al.* (1997b) Forcings and chaos in interannual to decadal climate change. *Journal of Geophysical Research*, **102**, 25679–25720.

Hansen, J.E., Sato, M., Lacis, A., Ruedy, R., Tegen, I. and Matthews, E. (1998) Climate forcings in the industrial era. *Proceedings of the National Academy of Sciences USA*, **95**, 12753–12758.

Harris, R.N. and Chapman, D.S. (1997) Borehole temperatures and a baseline for 20th-century global warming estimates. *Science*, **275**, 1618–1621.

Harvey, L.D.D. (1988a) Climatic impact of ice age aerosols. *Nature*, **334**, 333–335.

Harvey, L.D.D. (1989) Effect of model structure on the response of terrestrial biosphere models to CO_2 and temperature increases. *Global Biogeochemical Cycles*, **3**, 137–153.

Harvey, L.D.D. (1993) A guide to global warming potentials (GWPs). Energy Policy, **21**, 24–34.

Harvey, L.D.D. (1994) Transient temperature and sea level response of a two-dimensional ocean–climate model to greenhouse gas increases. *Journal of Geophysical Research*, **99**, 18447–18466.

Harvey, L.D.D. (1995) Impact of isopycnal diffusion on heat fluxes and the transient response of a two-dimensional ocean model. *Journal of Physical Oceanography*, **25**, 2166–2176.

Harvey, L.D.D. (1996a) Development of a risk-hedging CO_2-emission policy, Part I: Risks of unrestrained emissions. *Climatic Change*, **34**, 1–40.

Harvey, L.D.D. (1996b) Development of a risk-hedging CO_2-emission policy, Part II: Risks associated with measures to limit emissions, synthesis, and conclusions. *Climatic Change*, **34**, 41–71.

Harvey, L.D.D. (1999) *Global Warming: The Hard Science*. Pearson Education, Harlow.

Harvey, L.D.D. and Huang, Z. (1995) An evaluation of the potential impact of methane-clathrate destabilization on future global warming. *Journal of Geophysical Research*, **100**, 2905–2926.

Harvey, L.D.D. and Huang, Z. (1999) A quasi-one-dimensional coupled climate-carbon cycle model: description, calibration, and sensitivity. *Journal of Geophysical Research* (submitted).

Harvey, L.D.D., Gregory, J., Hoffert, M., Jain, A., Lal, M., Leemans, R., Raper, S., Wigley, T. and de Wolde, J. (1997) *An Introduction to Simple Climate Models Used in the IPCC Second Assessment Report*, Intergovernmental Panel on Climate Change, Technical Paper II, 50 pages.

Hauglustaine, D.A., Granier, C., Brasseur, G.P. and Mégie, G. (1994) The importance of atmospheric chemistry in the calculation of radiative forcing on the climate system. *Journal of Geophysical Research*, **99**, 1173–1186.

Haywood, J.M., Schwarzkopf, M.D. and Ramaswamy, V. (1998) Estimates of radiative forcing due to modeled increases in tropospheric ozone. *Journal of Geophysical Research*, **103**, 16999–17007.

Heimann, M. and Maier-Reimer, E. (1994) On the relations between the uptake of carbon dioxide and its isotopes by the ocean. *Global Biogeochemical Cycles*, **10**, 89–110.

Held, I.M. and Suarez, M.J. (1974) Simple albedo feedback models of the icecaps. *Tellus*, **26**, 613–630.

Henderson-Sellers, A. *et al.* (1998) Tropical cyclones and global climate change: a post-IPCC assessment. *Bulletin of the American Meteorological Society*, **79**, 19–38.

Hennessy, K.J., Gregory, J.M. and Mitchell, J.F.B. (1997) Changes in daily precipitation under enhanced greenhouse conditions. *Climate Dynamics*, **13**, 667–680.

Hense, A., Krahe, P. and Flohn, H. (1988) Recent fluctuations of tropospheric temperature and water vapour content in the tropics. *Meterology and Atmospheric
Physics*, **38**, 215–227.

Hirakuchi, H. and Giorgi, F. (1995) Multiyear present-day and $2 \times CO_2$ simulations
of monsoon climate over eastern Asia and Japan with a regional climate model
nested in a general circulation model. *Journal of Geophysical Research*, **100**,
21105–21125.

Hoffert, M.I. and Covey, C. (1992) Deriving global climate sensitivity from paleocli-
mate reconstructions. *Nature*, **360**, 573–576.

Hoffert, M.I., Callegari, A.J. and Hseih, C.-T. (1980) The role of deep sea heat storage
in the secular response to climatic forcing. *Journal of Geophysical Research*, **85**,
6667–6679.

Hoffert, M.I., Callegari, A.J. and Hseih, C.-T. (1981) A box-diffusion carbon cycle
model with upwelling, polar bottom water formation and a marine biosphere, in
Bolin, B. (ed.), *Carbon Cycle Modeling, SCOPE 16*, John Wiley and Sons, New
York, 287–305.

Hoffert, M.I. and Flannery, B.P. (1985) Model projections of the time-dependent
response to increasing carbon dioxide, in MacCraken, M.C. and Luther, F. M.
(eds), *Projecting the Climatic Effects of Increasing Carbon Dioxide*, DOE/ER–0237,
US Department of Energy, Washington, 149–190.

Hoffert, M.I., Frei, A. and Narayanan, V.K. (1988) Application of solar max acrim
data to analysis of solar-driven climatic variability on earth. *Climatic Change*, **13**,
267–286.

Hoffert, M.I. *et al.* (1998) Energy implications of future stabilization of atmospheric
CO_2 content. *Nature*, **395**, 881–884.

Houghton, R.A. (1987) Biotic changes consistent with the increased seasonal amplitude of atmospheric CO_2 concentrations. *Journal of Geophysical Research*, **92**,
4223–4230.

Houghton, R.A. (1996) Terrestrial sources and sinks of carbon inferred from terrestrial
data. *Tellus*, **48**, 420–433.

Houghton, R.A. (1998) The annual net flux of carbon to the atmosphere from
changes in land use, 1850–1990. *Tellus* (in press).

Hsieh, W.W. and Bryan, K. (1996) Redistribution of sea level rise associated with
enhanced greenhouse warming: a simple model study. *Climate Dynamics*, **12**,
535–544.

Hulme, M. (1991) An intercomparison of model and observed global precipitation
climatologies. *Geophysical Research Letters*, **18**, 1715–1718.

Hulme, M., Zhao, Z.-C. and Jiang, T. (1994) Recent and future climate change in
East Asia. *International Journal of Climatology*, **14**, 637–658.

Huybrechts, P. and de Wolde, J. (1999) The dynamic response of the Greenland
and Antarctic ice sheets to multiple-century climatic warming. *Journal of
Climate* (submitted).

Huybrechts, P. and Oerlemans, J. (1990) Response of the Antarctic ice sheet to future
greenhouse warming. *Climate Dynamics*, **5**, 93–102.

Huybrechts, P., Letreguilly, A. and Reeh, N. (1991) The Greenland ice sheet
and greenhouse warming. *Palaeogeography, Palaeoclimatology, Palaeoecology*, **89**,
399–412.

Idso, K.E. and Idso, S.B. (1994) Plant responses to atmospheric CO_2 enrichment in
the face of environmental constraints: a review of the past 10 years' research.
Agricultural and Forest Meteorology, **69**, 153–203.

IEA (International Energy Agency) (1991a) *Greenhouse Gases, Abatement and Control*, International Energy Agency, Paris.

IEA (International Energy Agency) (1991b) *Greenhouse Gas Emissions: The Energy Dimension*, International Energy Agency, Paris.

Ishitani, H. *et al.* (1996) Energy supply mitigation options, in Watson, R.T., Zinyowera, M.C. and Moss, R.H. (eds), *Climate Change 1995 – Impacts, Adaptations and Mitigation of Climate Change: Scientific–Technical Analysis*, Cambridge University Press, Cambridge, 587–647.

Jaffe, D.A. (1992) The nitrogen cycle, in Butcher, S.S., Charlson, R.J., Orians, G.H. and Wolfe, G.V. (eds), *Global Biogeochemical Cycles*, Academic Press, London, 263–284.

Jain, A.K., Kheshgi, H.S., Hoffert, M.I. and Wuebbles, D.J. (1995) Distribution of radiocarbon as a test of global carbon cycle models. *Global Biogeochemical Cycles*, **9**, 153–166.

Jaques, A.P. (1992) *Canada's Greenhouse Gase Emissions: Estimates for 1990*, Report EPS 5/AP/4, Environment Canada, Ottawa, 78 pages.

Jiang, Y. and Yung, Y.L. (1996) Concentrations of tropospheric ozone from 1979 to 1992 over tropical Pacific South America from TOMS data. *Science*, **272**, 714–716.

Johns, T.C., Carnell, R.E., Crossley, J.F., Gregory, J.M., Mitchell, J.F.B., Senior, C.A., Tett, S.F.B. and Wood, R.A. (1997) The second Hadley Centre coupled ocean–atmosphere GCM: model description, spinup and validation. *Climate Dynamics*, **13**, 103–134.

Johnsen, S.J., Dansgaard, W. and White, J.W.C. (1989) The origin of Arctic precipitation under present and glacial conditions. *Tellus*, **41B**, 452–468.

Jonas, P.R., Charlson, R.J. and Rodhe, H. (1995) Aerosols, in Houghton, J.T., Filho, L.G.M., Bruce, J., Lee, H., Callander, B.A., Haites, E., Harris, N. and Maskell, K. (eds), *Climate Change 1994: Radiative Forcing of Climate Change and an Evaluation of the IPCC IS92 Emission Scenarios*, Cambridge University Press, Cambridge, 127–162.

Jones, A., Roberts, D.L. and Slingo, A. (1994) A climate model study of indirect radiative forcing by anthropogenic sulphate aerosols. *Nature*, **370**, 450–453.

Jones, P.D. (1994) Hemisphere surface air temperature variations: a reanalysis and an update to 1993. *Journal of Climate*, 7, 1794–1802.

Jones, P.D., Wigley, T.M.L. and Wright, P.B. (1986) Global temperature variations between 1861–1984. *Nature*, **322**, 430–434.

Jones, P.D., Groisman, P. Ya., Coughlan, M., Plummer, N., Wang, W.-C. and Karl, T.R. (1990) Assessment of urbanization effects in time series of surface air temperature over land. *Nature*, **347**, 169–172.

Jones, P.D., Wigley, T.M.L. and Farmer, G. (1991) Marine and land temperature data sets: a comparison and a look at recent trends, in Schlesinger, M.E. (ed.), *Greenhouse-gas-induced Climatic Change: A Critical Appraisal of Simulations and Observations*, Developments in Atmospheric Science, **19**, Elsevier, Amsterdam, 153–172.

Jones, P.D., Osborn, T.J. and Briffa, K.R. (1997) Estimating sampling errors in large-scale temperature averages. *Journal of Climate*, **10**, 2548–2568.

Kane, R.P. (1997) Relationship of El Niño–Southern Oscillation and Pacific sea surface temperature with rainfall in various regions of the globe. *Monthly Weather Review*, **125**, 1792–1800.

Karl, T.R., Knight, R.W. and Plummer, N. (1995) Trends in high-frequency climate variability in the twentieth century. *Nature*, **377**, 217–220.

Kashiwagi, T. *et al.* (1996) Industry, in Watson, R.T., Zinyowera, M.C. and Moss, R.H. (eds), *Climate Change 1995 – Impacts, Adaptations and Mitigation of Climate Change: Scientific–Technical Analysis*, Cambridge University Press, Cambridge, 649–677.

Kattenberg, A., Giorgi, F., Grassl, H., Meehl, G.A., Mitchell, J.F.B., Stouffer, R.J., Tokioka, T., Weaver, A.J. and Wigley, T.M.L. (1996) Climate models – projections of future climate, in Houghton, J.T., Filho, L.G.F., Callander, B.A., Harris, N., Kattenberg, A. and Maskell, K. (eds), *Climate Change 1995: The Science of Climate Change*, Cambridge University Press, Cambridge, 285–357.

Keeling, C.D. and Whorf, T.P. (1998) Atmospheric CO_2 records from sites in the SIO air sampling network, in Trends: *A Compendium of Data on Global Change*, Carbon Dioxide Information Analysis Center, Oak Ridge National Laboratory, Oak Ridge, Tennessee.

Keeling, C.D., Whorf, T.P., Wahlen, M. and van der Plicht, J. (1995) Interannual extremes in the rate of rise of atmospheric carbon dioxide since 1980. *Nature*, **375**, 666–670.

Keeling, R.F., Piper, S.C. and Heimann, M. (1996) Global and hemispheric CO_2 sinks deduced from changes in atmospheric O_2 concentration. *Nature*, **381**, 218–221.

Keller, M., Jacob, D.J., Wofsy, S.C. and Harris, R.C. (1991) Effects of tropical deforestation on global and regional atmospheric chemistry. *Climatic Change*, **19**, 139–158.

Kheshgi, H.S. and Lapenis, A.G. (1996) Estimating the accuracy of Russian paleo-temperature reconstructions. *Palaeogeography, Palaeoclimatology, Palaeoecology*, **121**, 221–237.

Kiladis, G.N. and Diaz, H.F. (1989) Global climatic anomalies associated with extremes in the Southern Oscillation. *Journal of Climate*, **2**, 1069–1090.

Kirchgessner, D.A., Piccot, S.D. and Winkler, J.D. (1993) Estimate of global methane emissions from coal mines. *Chemosphere*, **26**, 453–472.

Knutson, T.R., Manabe, S. and Gu, D. (1997) Simulated ENSO in a global coupled ocean–atmosphere model: multidecadal amplitude modulation and CO_2 sensitivity. *Journal of Climate*, **10**, 138–161.

Knutson, T.R., Tuleya, R.E. and Kurihara, Y. (1998) Simulated increase of hurricane intensities in a CO_2-warmed climate. *Science*, **279**, 1018–1020.

Ko, M.K.W., Sze, N.D., Wang, W.-C., Shia, G., Goldman, A., Murcray, F.J., Murcray, D.G. and Rinsland, C.P. (1993) Atmospheric sulfur hexafluoride: sources, sinks, and greenhouse warming. *Journal of Geophysical Research*, **98**, 10499–10507.

Krol, M., van Leeuwen, P.J. and Lelieveld, J. (1998) Global OH trend inferred from methylchloroform measurements. *Journal of Geophysical Research*, **103**, 10697–10711.

Lal, M. and Ramanathan, V. (1984) The effects of moist convection and water vapor processes on climate sensitivity. *Journal of the Atmospheric Sciences*, **41**, 2238–2249.

Lambert, S.J. (1996) Intense extratropical northern hemisphere winter cyclone events: 1899–1991. *Journal of Geophysical Research*, **101**, 21319–21325.

Landsea, C.W., Nicholls, N., Gray, W.M. and Avila, L.A. (1996) Downward trends in the frequency of intense Atlantic hurricanes during the past five decades. *Geophysical Research Letters*, **23**, 1697–1700.

Lapenis, A.G. and Shabalova, M.V. (1994) Global climate changes and moisture conditions in the intracontinental arid zones. *Climatic Change*, **27**, 259–297.

Law, K.S. and Nisbet, E.G. (1996) Sensitivity of the CH_4 growth rate to changes in CH4 emissions from natural gas and coal. *Journal of Geophysical Research*, **101**, 14387–14397.

Le Treut, H. and Li, Z.-X. (1991) Sensitivity of an atmospheric general circulation model to prescribed SST changes: feedback effects associated with the simulation of cloud optical properties. *Climate Dynamics*, **5**, 175–187.

Le Treut, H., Li, Z.X. and Forichon, M. (1994) Sensitivity of the LMD general circulation model to greenhouse forcing associated with two different cloud water parameterizations. *Journal of Climate*, 7, 1827–1841.

Lean, J., Beer, J. and Bradley, R. (1995) Reconstruction of solar irradiance since 1610: implications for climate change. *Geophysical Research Letters*, **22**, 3195–3198.

Leggett, J., Pepper, W.J. and Swart, R.J. (1992) Emissions scenarios for the IPCC: an update, in Houghton, J.T., Callander, B.A. and Varney, S.K. (eds), *Climate Change 1992: The Supplementary Report to the IPCC Scientific Assessment*, Cambridge University Press, Cambridge, 69–95.

Lelieveld, J. and Crutzen, P.J. (1992) Indirect chemical effects of methane on climate warming. *Nature*, **355**, 339–341.

Lelieveld, J., Crutzen, P.J. and Dentener, F.J. (1998) Changing concentration, lifetime and climate forcing of atmospheric methane. *Tellus*, **50B**, 128–150.

Letréguilly, A., Huybrechts, P. and Reeh, N. (1991) Steady-state characteristics of the Greenland ice sheet under different climates. *Journal of Glaciology*, **37**, 149–157.

Levine, M.D. *et al.* (1996) Mitigation options for human settlements, in Watson, R.T., Zinyowera, M.C. and Moss, R.H. (eds), *Climate Change 1995 – Impacts, Adaptations and Mitigation of Climate Change: Scientific–Technical Analysis*, Cambridge University Press, Cambridge, 713–743.

Levitus, S. (1982) *Climatological Atlas of the World Ocean*, NOAA Professional Paper 13, Washington, DC, 173 pages.

Lindzen, R. (1990) Some coolness concerning global warming. *Bulletin of the American Meteorological Society*, **71**, 288–299.

Lindzen, R. and Giannitsis, C. (1998) On the climatic implications of volcanic cooling. *Journal of Geophysical Research*, **103**, 5929–5941.

Liousse, C., Penner, J.E., Chuang, C., Walton, J.J., Eddleman, H. and Cachier, H. (1996) A global three-dimensional model study of carbonaceous aerosols. *Journal of Geophysical Research*, **101**, 19411–19432.

Logan, J.A. (1994) Trends in the vertical distribution of ozone: an analysis of ozonesonde data. *Journal of Geophysical Research*, **99**, 25553–25585.

Lohmann, U. and Feichter, J. (1997) Impact of sulfate aerosols on albedo and lifetime of clouds: a sensitivity study with the ECHAM4 GCM. *Journal of Geophysical Research*, **102**, 13685–13700.

Lohmann, U. and Roeckner, E. (1995) Influence of cirrus cloud radiative forcing on climate and climate sensitivity in a general circulation model. *Journal of Geophysical Research*, **100**, 16305–16323.

Luo, Y., Jackson, R.B., Field, C.B. and Mooney, H.A. (1996) Elevated CO_2 increases belowground respiration in California grasslands. *Oecologia*, **108**, 130–137.

Machida, T., Nakazawa, T., Fujii, Y., Aoki, S. and Watanabe, O. (1995) Increase in the atmospheric nitrous oxide concentration during the last 250 years. *Geophysical Research Letters*, **22**, 2921–2924.

Maier-Reimer, E., Mikolajewicz, U. and Winguth, A. (1996) Future ocean uptake of CO_2: interactions between ocean circulation and biology. *Climate Dynamics*, **12**, 711–721.

Manabe, S. and Bryan, K. (1985) CO_2-induced change in a coupled ocean–atmosphere model and its paleoclimatic implications. *Journal of Geophysical Research*, **90**, 11689–11707.

Manabe, S. and Stouffer, R.J. (1994) Multiple-century response of a coupled ocean–atmosphere model to an increase of atmospheric carbon dioxide. *Journal of Climate*, 7, 5–23.

Manabe, S., Stouffer, R.J., Spelman, M.J. and Bryan, K. (1991) Transient responses of a coupled ocean–atmosphere model to gradual changes of atmospheric CO_2. Part I: Annual mean response. *Journal of Climate*, 4, 785–818.

Manabe, S., Spelman, M.J. and Stouffer, R.J. (1992) Transient responses of a coupled ocean–atmosphere model to gradual changes of atmospheric CO_2. Part II: Seasonal response. *Journal of Climate*, 5, 105–126.

Mann, M.E., Bradley, R.S. and Hughes, M.K. (1998) Global-scale temperature patterns and climate forcing over the past six centuries. *Nature*, 392, 779–787.

Marenco, A., Gouget, H., Nédélec, P., Pagés, J.P. and Karcher, F. (1994) Evidence of a long-term increase in tropospheric ozone from the Pic du Midi series: consequences: positive radiative forcing. *Journal of Geophysical Research*, 99, 16617–16632.

Marland, G., Boden, T.A., Andres, R.J., Brenkert, A.L. and Johnston, C. (1998) Global, regional, and national CO_2 emissions, in *Trends: A Compendium of Data on Global Change*, Carbon Dioxide Information Analysis Center, Oak Ridge National Laboratory, Oak Ridge, Tennessee.

Maslanik, J.A., Serreze, M.C. and Barry, R.G. (1996) Recent decreases in Arctic summer ice cover and linkages to atmospheric circulation changes. *Geophysical Research Letters*, 23, 1677–1680.

Mauzerall, D.L. *et al.* (1998) Photochemistry of biomass burning plumes and implications for tropospheric ozone over the tropical South Atlantic. *Journal of Geophysical Research*, 103, 8401–8423.

McConnaughay, K.D.M., Bassow, S.L., Berntson, G.M. and Bazzaz, F.A. (1996) Leaf senescence and decline of end-of-season gas exchange in five temperate deciduous tree species grown in elevated CO_2 concentrations. *Global Change Biology*, 2, 25–33.

Meehl, G.A. and Washington, W.M. (1988) A comparison of soil-moisture sensitivity in two global climate models. *Journal of the Atmospheric Sciences*, 45, 1476–1492.

Meehl, G.A. and Washington, W.M. (1993) South Asian summer monsoon variability in a model with doubled atmospheric carbon dioxide concentration. *Science*, 260, 1101–1103.

Meehl, G.A., Washington, W.M. and Karl, T. (1993) Low-frequency variability and CO_2 transient climate change. Part 1. Time-averaged differences. *Climate Dynamics*, 8, 117–133.

Meehl, G.A., Washington, W.M., Erickson III, D.J., Briegleb, B.P. and Jaumann, P.J. (1996) Climate change from increased CO_2 and direct and indirect effects of sulfate aerosols. *Geophysical Research Letters*, 23, 3755–3758.

Meehl, G.A., Boer, G.J., Covey, C., Latif, M. and Stouffer, R.J. (1997) Intercomparison makes for a better climate model. *EOS*, 78, 445–451.

Melillo, J.M., Prentice, I.C., Farquhar, G.D., Schulze, E.-D. and Sala, O.E. (1996) Terrestrial biotic responses to environmental change and feedbacks to climate, in Houghton, J.T., Filho, L.G.F., Callander, B.A., Harris, N., Kattenberg, A. and Maskell, K. (eds), *Climate Change 1995: The Science of Climate Change*, Cambridge University Press, Cambridge, 445–481.

Mercer, J.H. (1978) West Antarctic ice sheet and CO_2 greenhouse effect: a threat of disaster. *Nature*, 271, 321–325.

Michaelis, L. *et al.* (1996) Mitigation options in the transportation sector, in Watson, R.T., Zinyowera, M.C. and Moss, R.H. (eds), Climate *Change 1995 – Impacts, Adaptations and Mitigation of Climate Change: Scientific-Technical Analysis*, Cambridge University Press, Cambridge, 679–712.

Mikolajewicz, U., Santer, B.D. and Maier-Reimer, E. (1990) Ocean response to greenhouse warming. *Nature*, **345**, 589–593.

Miller, K.G., Fairbanks, R.G. and Mountain, G.S. (1987) Tertiary oxygen isotope synthesis, sea level history, and continental margin erosion. *Paleoceanography*, **2**, 1–19.

Minschwaner, K., Carver, R.W., Briegleb, B.P. and Roche, A.E. (1998) Infrared radiative forcing and atmospheric lifetimes of trace species based on observations from UARS. *Journal of Geophysical Research*, **103**, 23243–23253.

Misra, P.K., Bloxam, R., Fung, C. and Wong, S. (1989) Non-linear response of wet deposition to emissions reduction: a model study. *Atmospheric Environment*, **23**, 671–687.

Mitchell, J.F.B. and Ingram, W.J. (1992) Carbon dioxide and climate: mechanisms of changes in cloud. *Journal of Climate*, **5**, 5–21.

Mitchell, J.F.B. and Johns, T.C. (1997) On modifications of global warming by sulfate aerosols. *Journal of Climate*, **10**, 245–267.

Mitchell, J.F.B. and Warrilow, D.A. (1987) Summer dryness in northern mid-latitudes due to increased CO_2. *Nature*, **330**, 238–240.

Mitchell, J.F.B., Manabe, S., Meleshko, V. and Tokioka, T. (1990) Equilibrium climate change – and its implications for the future, in Houghton, J.T., Jenkins, G.J. and Ephraums, J.J. (eds), *Climate Change: The IPCC Scientific Assessment*, Cambridge University Press, Cambridge, 131–172.

Mohan, M. and Kumar, S. (1998) Review of acid rain potential in India: future threats and remedial measures. *Current Science*, **75**, 579–593.

Mosier, A.R., Duxbury, J.M., Freney, J.R., Heinemeyer, O., Kinami, K. and Johnson, D.E. (1998a) Mitigating agricultural emissions of methane. *Climatic Change*, **40**, 39–80.

Mosier, A.R., Duxbury, J.M., Freney, J.R., Heinemeyer, O. and Kinami, K. (1998b) Assessing and mitigating N_2O emissions from agricultural soils. *Climatic Change*, **40**, 7–38.

Murphy, J.M. and Mitchell, J.F.B. (1995) Transient response of the Hadley Centre coupled ocean–atmosphere model to increasing carbon dioxide. Part II: Spatial and temporal structure of response. *Journal of Climate*, **8**, 57–80.

Myers, N. (1989) *Deforestation Rates in Tropical Forests and their Climatic Implications*, Friends of the Earth, London.

Myhre, G. and Stordal, F. (1997) Role of spatial and temporal variations in the computation of radiative forcing and GWP. *Journal of Geophysical Research*, **192**, 11181–11200.

Myneni, R.B., Keeling, C.D., Tucker, C.J., Asrar, G. and Nemani, R.R. (1997) Increased plant growth in the northern high latitudes from 1981–1991. *Nature*, **386**, 698–702.

Mysak, L.A. and Manak, D.K. (1989) Arctic sea-ice extent and anomalies, 1953–1984. *Atmosphere–Ocean*, **27**, 376–405.

Najjar, R.G., Sarmiento, J.L. and Toggweiler, J.R. (1992) Downward transport and fate of organic matter in the ocean: simulations with a general circulation model. *Global Biogeochemical Cycles*, **6**, 45–76.

Nakicenovic, N., Grübler, A., Ishitani, H., Johansson, T., Marland, G., Moreira, J.R. and Rogner, H.-H. (1996) Energy primer, in Watson, R.T., Zinyowera, M.C., Moss, R.H. and Dokken, D.J. (eds), *Climate Change 1995 – Impacts, Adaptations and Mitigation of Climate Change: Scientific–Technical Analysis*, Cambridge University Press, Cambridge, 75–92.

Nedoluha, G.E., Bevilacqua, R.M., Gomez, R.M., Siskind, D.E., Hicks, B.C., Russell III, J.M. and Connor, B.J. (1998) Increases in middle atmospheric water vapor as observed by the Halogen Occultation Experiment and the ground-based Water Vapor Millimeter-wave Spectrometer from 1991 to 1997. *Journal of Geophysical Research*, **103**, 3531–3543.

Newell, R.E. (1978) The Global Circulation of the Atmosphere, Royal Meteorological Society, Bracknell, England.

Newman, M.J. and Rood, R.T. (1977) Implications of solar evolution for the Earth's early atmosphere. *Science*, **198**, 1035–1037.

Nicholls, K.W. (1997) Predicted reduction in basal melt rates of an Antarctic ice shelf in a warmer climate. *Nature*, **388**, 460–462.

Nicholls, N., Gruza, G.V., Jouzel, J., Karl, T.R., Ogallo, L.A. and Parker, D.E. (1996) Observed climate variability and change, in Houghton, J.T., Filho, L.G.F., Callander, B.A., Harris, N., Kattenberg, A. and Maskell, K. (eds), *Climate Change 1995: The Science of Climate Change*, Cambridge University Press, Cambridge, 133–192.

Norby, R.J., Gunderson, C.A., Wullschleger, S.D., O'Neill, E.G., and MacCracken, M.K. (1992) Productivity and compensatory response of yellow-poplar trees in elevated CO_2. *Nature*, **357**, 322–324.

Novelli, P.C., Masarie, K.A. and Lang, P.M. (1998) Distributions and recent changes of carbon monoxide in the lower troposphere. *Journal of Geophysical Research*, **103**, 19015–19033.

Oechel, W.C., Hastings, S.J., Vourlitis, G., Jenkins, M., Riechers, G. and Grulke, N. (1993) Recent change of Arctic tundra ecosystems from a net carbon dioxide sink to a source. *Nature*, **361**, 520–523.

Oerlemans, J. (1994) Quantifying global warming from retreat of glaciers. *Science*, **264**, 243–245.

Ohmura, A., Wild, M. and Bengtsson, L. (1996) A possible change in mass balance of Greenland and Antarctic ice sheets in the coming century. *Journal of Climate*, **9**, 2124–2135.

Oltmans, S.J. and Hofmann, D.J. (1995) Increase in lower-stratospheric water vapour at a mid-latitude Northern Hemisphere site from 1981 to 1994. *Nature*, **374**, 146–149.

Oort, A.H. and Liu, H. (1993) Upper-air temperature trends over the globe, 1958–1989. *Journal of Climate*, **6**, 292–307.

Oppenheimer, M. (1998) Global warming and the stability of the West Antarctic Ice Sheet. *Nature*, **393**, 325–332.

Orr, J.C. (1993) Accord between ocean models predicting uptake of anthropogenic CO_2. *Water, Air, and Soil Pollution*, **70**, 465–481.

Osborn, T.J. and Wigley, T.M.L. (1994) A simple model for estimating methane concentrations and lifetime variations. *Climate Dynamics*, **9**, 181–193.

Parker, D.E. and Cox, D.I. (1995) Towards a consistent global climatological rawinsonde data-base. *International Journal of Climatology*, **15**, 473–496.

Parker, D.E., Folland, C.K. and Jackson, M. (1995) Marine surface temperature: observed variations and data requirements. *Climatic Change*, **31**, 559–600.

Pastor, J. and Post, W.M. (1988) Response of northern forests to CO_2-induced climate change. *Nature*, **334**, 55–58.

Pearce, D.W., Cline, W.R., Achanta, A.N., Fankhauser, S., Pachauri, R.K., Tol, R.S.J. and Vellinga, P. (1996) The social costs of climate change: greenhouse damage and the benefits of control, in Bruce, J.P., Lee, H. and Haites, E.F. (eds), *Climate*

Change 1995 – Economic and Social Dimensions of Climate Change, Cambridge University Press, Cambridge, 179–224.

Peltier, W.R. and Tushingham, A.M. (1989) Global sea level rise and the greenhouse effect: might they be connected? *Science*, **244**, 806–810

Pham, M., Müller, J.-F., Brasseur, G.P., Granier, C. and Mégie, G. (1996) A 3D model study of the global sulphur cycle: contributions of anthropogenic and biogenic sources. *Atmospheric Environment*, **30**, 1815–1822.

Plöchl, M. and Cramer, W. (1995) Coupling global models of vegetation structure and ecosystem processes. *Tellus*, **47B**, 240–250.

Pollard, D. and Thompson, S.L. (1994) Sea-ice dynamics and CO_2 sensitivity in a global climate model. *Atmosphere–Ocean*, **32**, 449–467.

Portmann, R.W., Solomon, S., Fishman, J., Olson, J.R., Kiehl, J.T. and Briegleb, B. (1997) Radiative forcing of the Earth's climate system due to tropical tropospheric ozone production. *Journal of Geophysical Research*, **102**, 9409–9417.

Post, W.M., Pastor, J., King, A.W. and Emanuel, W.R. (1992) Aspects of the interaction between vegetation and soil under global change. *Water, Air, and Soil Pollution*, **64**, 345–363.

Prather, M., Derwent, R., Ehhalt, D., Fraser, P., Sanhueza, E. and Zhou, X. (1995) Other trace gases and atmospheric chemistry, in Houghton, J.T., Filho, L.G.M., Bruce, J., Lee, H., Callander, B.A., Haites, E., Harris, N. and Maskell, K. (eds), *Climate Change 1994: Radiative Forcing of Climate Change and an Evaluation of the IPCC IS92 Emission Scenarios*, Cambridge University Press, Cambridge, 73–126.

Princiotta, F.T. (1991) Pollution control for utility power generation, 1990–2020, in Tester, J.W., Wood, D.O. and Ferrari, N.A. (eds), Energy and the Environment in the 21st Century, MIT Press, Cambridge, MA, 631–650.

Prinn, R.G., Weiss, R.F., Miller, B.R., Huang, J., Alyea, F.N., Cunnold, D.M., Fraser, P.J., Hartley, D.E. and Simmonds, P.G. (1995) Atmospheric trends and lifetime of CH3CCl3 and global OH concentrations. *Science*, **269**, 187–192.

Qian, J.J. and Zhang, K.M. (1998) China's desulfurization potential. *Energy Policy*, **26**, 345–351.

Ramaswamy, V., Schwarzkopf, M.D. and Randel, W.J. (1996) Fingerprint of ozone depletion in the spatial and temporal pattern of recent lower-stratospheric cooling. *Nature*, **382**, 616–618.

Randall, D.A. *et al.* (1994) Analysis of snow feedbacks in 14 general circulation models. *Journal of Geophysical Research*, **99**, 20757–20771.

Rasch, P.J., Barth, M.C., Kiehl, J.T., Schwartz, S.E. and Benkowitz, C.M. (1999) A description of the global sulphur cycle and its controlling processes in the NCAR CCM3. *Journal of Geophysical Research* (submitted).

Repetto, R. and Austin, D. (1997) *The Costs of Climate Protection: A Guide for the Perplexed*, World Resources Institute, Washington, 51 pages.

Rind, D. (1987) The doubled CO_2 climate: impact of the sea surface temperature gradient. *Journal of the Atmospheric Sciences*, **44**, 3235–3268.

Rind, D. (1988) The doubled CO_2 climate and the sensitivity of the modeled hydrologic cycle. *Journal of Geophysical Research*, **93**, 5385–5412.

Rind, D., Goldberg, R., Hansen, J., Rosenzweig, C. and Ruedy, R. (1990) Potential evapotranspiration and the likelihood of future drought. *Journal of Geophysical Research*, **95**, 9983–10004.

Rind, D., Healy, R., Parkinson, C. and Martinson, D. (1995) The role of sea ice in $2 \times CO_2$ climate model sensitivity. Part I: The total influence of sea ice thickness and extent. *Journal of Climate*, **8**, 449–463.

Robock, A. and Free, M.P. (1995) Ice cores as an index of global volcanism from 1850 to the present. *Journal of Geophysical Research*, **100**, 11549–11567.

Robock, A. and Mao, J. (1995) The volcanic signal in surface temperature observations. *Journal of Climate*, **8**, 1086–1103.

Roelofs, G.-J., Lelieveld, J. and Ganzeveld, L. (1998) Simulation of global sulfate distribution and the influence on effective cloud drop radii with a coupled photochemistry–sulfur cycle model. *Tellus*, **50B**, 224–242.

Rosa, L.P. and Schaeffer, R. (1995) Global warming potentials: the case of emissions from dams. *Energy Policy*, **23**, 149–158.

Rothman, D.S. (1999) Measuring environmental values and environmental impacts: going from the local to the global. *Climatic Change*, (in press).

Ryan, M.G. (1991) Effects of climate change on plant respiration. *Ecological Applications*, **1**, 157–167.

Sagan, C., Toon, O.B. and Pollak, J.B. (1979) Anthropogenic albedo changes and the Earth's climate. *Science*, **206**, 1363–1368.

Sahagian, D.L., Schwartz, F.W. and Jacobs, D.K. (1994) Direct anthropogenic contributions to sea level rise in the twentieth century. *Nature*, **367**, 54–57.

Sandroni, S., Anfossi, D. and Viarengo, S. (1992) Surface ozone levels at the end of the nineteenth century in South America. *Journal of Geophysical Research*, **97**, 2535–2539.

Santer, B.D. *et al.* (1996) A search for human influences on the thermal structure of the atmosphere. *Nature*, **382**, 39–46.

Sarmiento, J.L. and Le Quéré, C. (1996) Oceanic carbon dioxide uptake in a model of century-scale global warming. *Science*, **274**, 1346–1350.

Sarmiento, J.L. and Sundquist, E.T. (1992) Revised budget for the oceanic uptake of anthropogenic carbon dioxide. *Nature*, **356**, 589–593.

Sarmiento, J.L., Orr, J.C. and Siegenthaler, U. (1992) A perturbation simulation of CO_2 uptake in an ocean general circulation model. *Journal of Geophysical Research*, **97**, 3621–3645.

Sarmiento, J.L., Slater, R.D., Fasham, M.J.R., Ducklow, H.W., Toggweiler, J.R. and Evans, G.T. (1993) A seasonal three-dimensional ecosystem model of nitrogen cycling in the North Atlantic euphotic zone. *Global Biogeochemical Cycles*, **7**, 417–450.

Sato, M., Hansen, J.E., McCormick, M.P. and Pollack, J.B. (1993) Stratospheric aerosol optical depths, 1850–1990. *Journal of Geophysical Research*, **98**, 22987–22994.

Schimel, D. *et al.* (1996) Radiative forcing of climate change, in Houghton, J.T., Filho, L.G.F., Callander, B.A., Harris, N., Kattenberg, A. and Maskell, K. (eds), *Climate Change 1995: The Science of Climate Change*, Cambridge University Press, Cambridge, 65–131.

Schlesinger, W.H. (1991) *Biogeochemistry: An Analysis of Global Change*, First Edition, Academic Press, San Diego, 443 pages.

Schlesinger, W.H. (1997) *Biogeochemistry: An Analysis of Global Change*, Second Edition, Academic Press, San Diego, 588 pages.

Schult, I., Feichter, J. and Cooke, W.F. (1997) Effect of black carbon and sulfate aerosols on the global radiation budget. *Journal of Geophysical Research*, **102**, 30107–30117.

Sebenius, J.K. (1991) Designing negotiations toward a new regime: the case of global warming. *International Security*, **15**, 110–148.

Seitz, F. (1994) *Global Warming and Ozone Hole Controversies: A Challenge to Scientific Judgment*, George C. Marshall Institute, Washington, 34 pages.

Sellers, P.J. *et al.* (1996) A revised land surface parameterization (SiB2) for atmospheric GCMs. Part I: Model formulation. *Journal of Climate*, **9**, 676–705.

Senior, C.A. and Mitchell, J.F.B. (1993) Carbon dioxide and climate: the impact of cloud parameterization. *Journal of Climate*, **6**, 393–418.

Shindell, D.T., Rind, D. and Lonergan, P. (1998) Increased polar stratospheric ozone losses and delayed eventual recovery owing to increasing greenhouse-gas concentrations. *Nature*, **392**, 589–592,

Shine, K.P., Derwent, R.G., Wuebbles, D.J. and Morcrette, J.-J. (1990) Radiative forcing of climate, in Houghton, J.T., Jenkins, G.J. and Ephraums, J.J. (eds), *Climate Change: The IPCC Scientific Assessment*, Cambridge University Press, Cambridge, 41–68.

Shine, K.P., Fouquart, Y., Ramaswamy, V., Solomon, S. and Srinivasan, J. (1995) Radiative forcing, in Houghton, J.T., Filho, L.G.M., Bruce, J., Lee, H., Callander, B.A., Haites, E., Harris, N. and Maskell, K. (eds), *Climate Change 1994: Radiative Forcing of Climate Change and an Evaluation of the IPCC IS92 Emission Scenarios*, Cambridge University Press, Cambridge, 163–203.

Siegenthaler, U. and Joos, F. (1992) Use of a simple model for studying oceanic tracer distributions and the global carbon cycle. *Tellus*, **44B**, 186–207.

Siegenthaler, U. and Oeschger, H. (1987) Biospheric CO_2 emissions during the past 200 years reconstructed from deconvolution of ice core data. *Tellus*, **39B**, 140–154.

Siegenthaler, U. and Sarmiento, J.L. (1993) Atmospheric carbon dioxide and the ocean. *Nature*, **365**, 119–125.

Smith, D.M. (1998) Recent increase in the length of the melt season of perennial Arctic sea ice. *Geophysical Research Letters*, **25**, 655–658.

Smith, I.N., Dix, M. and Allan, R.J. (1997) The effect of greenhouse SSTs on ENSO simulations with an AGCM. *Journal of Climate*, **10**, 342–352.

Soden, B.J. and Fu, R. (1995) A satellite analysis of deep convection, upper-tropospheric humidity, and the greenhouse effect. *Journal of Climate*, **8**, 2333–2351.

Soden, B.J. (1997) Variations in the tropical greenhouse effect during El Niño. *Journal of Climate*, **10**, 1050–1055.

Solanki, S.K. and Fligge, M. (1998) Solar irradiance since 1874 revisited. *Geophysical Research Letters*, **25**, 341–344.

Solomon, A.M. (1986) Transient response of forests to CO_2-induced climate change: simulation modeling experiments in eastern North America. *Oecologia*, **68**, 567–579.

Solomon, S. and Daniel, J.S. (1996) Impact of the Montreal Protocol and its amendments on the rate of change of global radiative forcing. *Climatic Change*, **32**, 7–17.

Somerville, R. and Remer, L. (1984) Cloud optical thickness feedbacks in the CO_2 climate problem. *Journal of Geophysical Research*, **89**, 9668–9672.

Spiro, P.A., Jacob, D.J. and Logan, J.A. (1992) Global inventory of sulfur emissions with 1° × 1° resolution. *Journal of Geophysical Research*, **97**, 6023–6036.

Stammerjohn, S.E. and Smith, R.C. (1997) Opposing Southern Ocean climate patterns as revealed by trends in regional sea ice coverage. *Climatic Change*, **37**, 617–639.

Stephenson, D.B. and Held, I.M. (1993) GCM response of northern winter stationary waves and storm tracks to increasing amounts of carbon dioxide. *Journal of Climate*, **6**, 1859–1870.

Stern, D.I. and Kaufmann, R.K. (1996a) Estimates of global anthropogenic methane emissions 1860–1993. *Chemosphere*, **33**, 159–176.

Stern, D.I. and Kaufmann, R.K. (1996b) *Estimates of Global Anthropogenic Sulfate Emissions 1860–1993*, CEES Working Paper 9602, Center for Energy and Environmental Studies, Boston University, Boston, MA.

Stocker, T.F., Broecker, W.S. and Wright, D.G. (1994) Carbon uptake experiment with a zonally averaged global ocean circulation model. *Tellus*, **46B**, 103–122.

Suhre, K., Andreae, M.O. and Rosset, R. (1995) Biogenic sulfur emissions and aerosols over the tropical South Atlantic 2. One-dimensional simulation of sulfur chemistry in the marine boundary layer. *Journal of Geophysical Research*, **100**, 11323–11334.

Sun, D.-Z. and Lindzen, R.S. (1993) Distribution of tropical tropospheric water vapour. *Journal of the Atmospheric Sciences*, **50**, 1643–1660.

Sun, D.Z. and Oort, A.H. (1995) Humidity–temperature relationships in the tropical troposphere. *Journal of Climate*, **8**, 1974–1987.

Suppiah, R. and Hennessy, K.J. (1998) Trends in total rainfall, heavy rain events and number of dry days in Australia, 1910–1990. *International Journal of Climatology*, **18**, 1141–1164.

Syncrude Canada (1998) *Environmental Progress Summary*, http://www.syncrude.com/5_env/progress_98/so2.htm.

Takahashi, T., Broecker, W.S. and Bainbridge, A.E. (1981) The alkalinity and total carbon dioxide concentration in the world oceans, in Bolin, B. (ed.), *Carbon Cycle Modelling*, SCOPE 16, John Wiley, Chichester, 271–286.

Tans, P.P., Fung, I.Y. and Takahashi, T. (1990) Observational constraints on the global atmospheric CO$_2$ budget. *Science*, **247**, 1431–1438.

Tett, S.F.B. (1995) Simulation of El Niño–Southern Oscillation-like variability in a global AOGCM and its response to CO$_2$ increase. *Journal of Climate*, **8**, 1473–1502.

Trenberth, K.E. and Hoar, T.J. (1996) The 1990–1995 El Niño–Southern Oscillation event: longest on record. *Geophysical Research Letters*, **23**, 57–60.

Trenberth, K.E. and Hoar, T.J. (1997) El Niño and climatic change. *Geophysical Research Letters*, **24**, 3057–3060.

Trenberth, K.E., Houghton, J.T. and Meira Filho, L.G. (1996) The climate system: an overview, in Houghton, J.T., Filho, L.G.F., Callander, B.A., Harris, N., Kattenberg, A. and Maskell, K. (eds), *Climate Change 1995: The Science of Climate Change*, Cambridge University Press, Cambridge, 51–64.

van Dorland, R., Dentener, F.J. and Lelieveld, J. (1997) Radiative forcing due to tropospheric ozone and sulfate aerosols. *Journal of Geophysical Research*, **102**, 28079–28100.

Vaughan, D.G. and Doake, C.S.M. (1996) Recent atmospheric warming and retreat of ice shelves on the Antarctic Peninsula. *Nature*, **379**, 328–331.

Verbitsky, M. and Saltzman, B. (1995) Behavior of the East Antarctic ice sheet as deduced from a coupled GCM/ice-sheet model. *Geophysical Research Letters*, **22**, 2913–2916.

Vose, R.S., Schmoyer, R.L., Steurer, P.M. Pterson, T.C., Heim, R., Karl, T.R. and Eischeid, J. (1992) *The Global Historical Climatology Network: Long-term Monthly Temperature, Precipitation, Sea Level Pressure, and Station Pressure Data*, Carbon Dioxide Information and Analysis Center, Oak Ridge National Laboratory, Oak Ridge, Tennessee, Report ORNL/CDIAC-53, NDP-041.

Walsh, J.E. (1995) Long-term observations for monitoring of the cryosphere. *Climatic Change*, **31**, 369–394.

Wang, Y., Jacob, D.J. and Logan, J.A. (1998a) Global simulation of tropospheric O_3–NO_x–hydrocarbon chemistry 1. Model formulation. *Journal of Geophysical Research*, **103**, 10713–10725.

Wang, Y., Logan, J.A. and Jacob, D.J. (1998b) Global simulation of tropospheric O_3–NO_x-hydrocarbon chemistry 2. Model evaluation and global ozone budget. *Journal of Geophysical Research*, **103**, 10727–10755.

Warrick, R.A., Le Provost, C., Meier, M.F., Oerlemans, J. and Woodworth, P.L. (1996) Changes in sea level, in Houghton, J.T., Filho, L.G.F., Callander, B.A., Harris, N., Kattenberg, A. and Maskell, K. (eds), *Climate Change 1995: The Science of Climate Change*, Cambridge University Press, Cambridge, 359–405.

Washington, W.M. and Meehl, G.A. (1989) Climate sensitivity due to increased CO_2: experiments with a coupled atmosphere and ocean general circulation model. *Climate Dynamics*, **4**, 1–38.

Washington, W.M. and Meehl, G.A. (1993) Greenhouse sensitivity experiments with penetrative cumulus convection and tropical cirrus albedo effects. *Climate Dynamics*, **8**, 211–223.

Watterson, I.G., O'Farrell, S.P. and Dix, M.R. (1997) Energy and water transport in climates simulated by a general circulation model that includes dynamic sea ice. *Journal of Geophysical Research*, **102**, 11027–11037.

Wentz, F.J. and Schabel, M. (1998) Effects of orbital decay on satellite-derived lower-tropospheric temperature trends. *Nature*, **394**, 661–664.

Wetherald, R.T. and Manabe, S. (1988) Cloud feedback processes in a general circulation model. *Journal of the Atmospheric Sciences*, **45**, 1397–1415.

Wetherald, R.T. and Manabe, S. (1995) The mechanisms of summer dryness induced by greenhouse warming. *Journal of Climate*, **8**, 3096–3108.

Whalen, S.C. and Reeburgh, W.S. (1990) Consumption of atmospheric methane by tundra soils. *Nature*, **346**, 160–162.

Whiting, G.J. and Chanton, J.P. (1993) Primary production control of methane emission from wetlands. *Nature*, **364**, 794–795.

Wigley, T.M.L. (1989) Possible climate change due to SO_2-derived cloud condensation nuclei. *Nature*, **339**, 365–367.

Wigley, T.M.L. (1991) Could reducing fossil-fuel emissions cause global warming? *Nature*, **349**, 503–506.

Wigley, T.M.L. and Raper, S.C.B. (1995) An heuristic model for sea level rise due to the melting of small glaciers. *Geophysical Research Letters*, **22**, 2749–2752.

Wigley, T.M.L., Richards, R. and Edmonds, J.A. (1996) Economic and environmental choices in the stabilization of atmospheric CO_2 concentrations. *Nature*, **379**, 242–245.

World Meteorological Organization (1998) *Scientific Assessment of Ozone Depletion: 1998*, World Meteorological Organization Global Ozone Research and Monitoring Project – Report No. 44, Geneva.

World Resources Institute (1998) *World Resources 1998–99*, World Resources Institute, Washington, DC, 369 pages.

Wright, D.G. and Stocker, T.F. (1991) A zonally averaged ocean model for the thermohaline circulation, I, Model development and flow dynamics. *Journal of Physical Oceanography*, **21**, 1713–1724.

Wullschleger, S.D., Ziska, L.H. and Bunce, J.A. (1994) Respiratory responses of higher plants to atmospheric CO_2 enrichment. *Physiologia Plantarum*, **90**, 221–229.

Zak, D.R., Pregitzer, K.S., Curtis, P.S., Teeri, J.A., Fogel, R. and Randlett, D.L. (1993) Elevated atmospheric CO_2 and feedback between carbon and nitrogen cycles. *Plant and Soil,* **151**, 105–117.

Zhang, Y. and Wang, W.-C. (1997) Model-simulated northern winter cyclone and anticyclone activity under a greenhouse warming scenario. *Journal of Climate,* **10**, 1616–634.

Zuo, Z. and Oerlemans, J. (1997) Contribution of glacier melt to sea-level rise since AD 1865: a regionally differentiated calculation. *Climate Dynamics,* **13**, 835–845.

Web sites used

Listed below are the Web sites from which data were downloaded for use in this book, or to which the reader can refer for up-to-date information.

Chapter 1

Paleoclimatic indicators from ice cores and marine sediments were obtained from the Paleoclimatology Website of the US National Oceanographic and Atmospheric Administration. The home page is http://www.ngdc.noaa.gov/paleo.

Chapter 3

National GHG emission inventories prepared under the United Nations Framework Convention on Climate Change can be obtained from http://www.unfccc.de/.

GHG emission estimates can also be obtained from the Pacific Institute for Studies in Development, Environment, and Security: http://www.globalchange.org/.

Information on sulphur emissions associated with various industrial activities can be obtained from the Energy Information Administration of the US Department of Energy (http://www/eia/doe/gov/cneaf/coal/cia/), the acid rain program of the US Environmental Protection Agency (http://www.epa.gov/acidrain/), the natural gas industry (http://natural-gas.com/environment/acid/html), and Syncrude Canada (http://www.syncrude.com/5 env/progress 98/so2.htm).

Data on variations through time in GHG concentrations and emissions can be obtained from the Carbon Dioxide Information and Analysis Center (CDIAC) of the US Department of Energy (http://cdiac.esd.ornl.gov).

Estimates of the historical variation of global sulphur emissions can be obtained from the Centre for Resource and Environment Studies of the Australian National University (http://cres.anu.edu.au/~dstern/anzsee/datasite.html).

Chapter 4

Grid point and hemispherically and globally averaged data on changes in sea surface or surface air temperature during the past century or so can be obtained from the Climate Research Institute of the University of East Anglia (http://www.cru.uea.ac.uk). Much of the same data can be obtained from the UK Meteorological Office (http://www.meto.gov.uk). An independent analysis of surface air temperature is available from the Goddard Institute for Space Studies (http://www.gis.nasa.gov).

The variation in tropospheric temperature as obtained from a carefully chosen set of 63 radiosonde stations can be obtained from the Carbon Dioxide Information and Analysis Center (http://cdiac.esd.ornl.gov/ftp/ndp008).

The variation in stratospheric and tropospheric temperature as determined from the satellite-based Microwave Sounding Unit (currently without correction for the drift in satellite altitude) can be obtained via FTP from 198.116.60.11 (wind.atmos.uah.edu).

Chapter 9

Data on AGCMs and AOGCMs can be obtained from the web site of the Climate Model Intercomparison Project (CMIP): http://www-pcmdi.11n1.gov/cmip. Results of transient climatic change simulations from seven (as of this writing) AOGCMs are available at the IPCC Data Distribution Centre (*http://ipcc-ddc.cru,uea.ac.uk*). The data for each model consist of monthly averaged, grid point values of daily mean, minimum, and maximum temperatures; precipitation; winds; and other variables for every month from 1860 to 2099. For each model, results are given for one control run, eight runs with GHG forcing only, and eight runs with GHG plus aerosol forcing.

Overall

In addition to the sites listed above, a good starting point for information (including recent research results) and data pertaining to climatic change is the Global Change Master Directory (*http://gcmd.gsfc.nasa.gov*). Another convenient starting point for information on climatic change is the Carbon Dioxide Information and Analysis Center (*http://cdiac.esd.ornl.gov*), mentioned above. Finally the United States Environmental Protection Agency global warming web site (*http://www.epa.gov/global warming*) contains links to about 100 climate-related web sites, including research centres, business associations, the environment ministries of some other governments, and some non-governmental organizations.

Index